Scott S Legge

OXFORD MONOGRAPHS ON BIOGEOGRAPHY

Editors: A. HALLAM, B. R. ROSEN, AND T. C. WHITMORE

OXFORD MONOGRAPHS ON
BIOGEOGRAPHY

Editors

A. Hallam, School of Earth Sciences, University
of Birmingham

B. R. Rosen, Department of Palaeontology,
National History Museum, London.

T. C. Whitmore, Department of Geography, University of Cambridge.

The Africa–South America Connection

Edited by

WILMA GEORGE

Formerly Lecturer in Zoology in the University of Oxford and
Fellow of Lady Margaret Hall, Oxford

and

RENÉ LAVOCAT

Directeur Honoraire de Laboratoire (Paléontologie des Vertébrés)
à l'École Pratique des Hautes Études
Université des Sciences et Techniques du Languedoc, Montpellier

CLARENDON PRESS · OXFORD
1993

Oxford University Press, Walton Street, Oxford OX2 6DP
Oxford New York Toronto
Delhi Bombay Calcutta Madras Karachi
Petaling Jaya Singapore Hong Kong Tokyo
Nairobi Dar es Salaam Cape Town
Melbourne Auckland
and associated companies in
Berlin Ibadan

Oxford is a trade mark of Oxford University Press

Published in the United States
by Oxford University Press, New York

Library of Congress Cataloging in Publication Data
The Africa–South America connection / edited by Wilma George and René Lavocat.
(Oxford monographs on biogeography : 7)
Includes bibliographic references and index
1. Biogeography—Africa. 2. Biogeography—South America.
3. Continental drift. 4. Paleogeography—Africa.
5. Paleogeography—South America. I. George, Wilma B. II. Lavocat, René.
III. Series: Oxford monographs on biogeography : no. 7.
QH194.A32 1993 574.96—dc20 92-27733
ISBN 0 19 8545770

Typeset by Joshua Associates Ltd, Oxford
Printed and bound in Great Britain by
BPCC Hazells Ltd, Aylesbury

PREFACE

The idea of this book, the plan, the first choice of authors, and the invitation to the contributors were entirely the result of the inspiration and the activity of Wilma George. I met her twice, at a symposium (No. 34) of the Zoological Society of London on the biology of the hystricomorph rodents, held in 1973, and at a symposium on the evolutionary relationships among rodents, held in Paris in 1985, at which she strongly supported my hypothesis about the relationship between the African and South American rodents.

A result of that convergence of opinion was that when she unfortunately died, leaving her work unfinished, I was asked to accept the responsibility for bringing it to completion. Although then already in my eighties, I felt that it was my duty to agree, as a friendly act of homage to her memory. Much of the draft material was becoming out of date because of delays caused by her final illness. I must heartily thank the authors, who willingly accepted the call to rewrite their contributions. Only minor additions were necessary for Wilma George's own chapter on the rodents, which we publish practically as she left it.

Wilma George evidently envisaged a broad treatment of the subject. In order to fulfil her intention, I have introduced new contributions accordingly.

To all the contributors, to everyone who helped me in this task, to Mr George Crowther, and to the Oxford University Press, who honoured me by offering this responsibility for a publication not in my own language in this prestigious series, I am extremely grateful. I hope that I have not proved too inadequate to my task, and that the result, with the help of so much good will, is a text that will be stimulating for the reader.

Teyran
December 1991   R. L.

CONTENTS

CONTRIBUTORS

Leslie C. Aiello: Department of Anthropology, University College London, Gower Street, London WC1E 6BT, UK.

C. Amedegnato: CNRS, Laboratoire d'Entomologie, Muséum National d'Histoire Naturelle, 45 rue Buffon, Paris 75005, France.

Eric Buffetaut: CNRS, Laboratoire de Paléontologie des Vertébrés, Université de Paris VI, 4 place Jussieu, 75252 Paris Cedex 05, France.

Marian Dawkins: Department of Zoology, South Parks Road, Oxford OX1 3PS, UK.

Wilma George: Formerly of Department of Zoology, South Parks Road, Oxford OX1 3PS, UK.

René Lavocat: 7 rue de l'Avenir, 34820 Teyran, France.

Simon J. Mayo: Herbarium, Royal Botanic Gardens, Kew, Richmond, Surrey TW9 3AE, UK.

Judith Totman Parrish: Department of Geosciences, Gould–Simpson Building, University of Arizona, Tucson, Arizona 85721, USA.

Jean-Claude Rage: CNRS, Laboratoire de Paléontologie des Vertébrés, Université de Paris VI, 4 place Jussieu, 75252 Paris Cedex 05, France.

Gerhard Storch: Forschungsinstitut Senckenberg, Senckenberg-Anlage 25, D-6000 Frankfurt am Main, Germany.

Albert E. Wood: 20 East Mechanic Street, Cape May Court House, NJ 08210, USA.

Wilma George explores the rock shelter of a noki, *Petrodromous typicus* in Augrabe Falls National Park, South Africa

WILMA GEORGE (Wilma Crowther)

My first sight of Wilma Crowther was across a sunny room in Lady Margaret Hall, Oxford, one afternoon in 1963. I saw a small woman with short, straight hair sitting in a cane chair. Behind her was a completely black wall on which a life-sized picture of a gazelle in the style of a cave painting had been expertly drawn in white paint. On her hand she wore a large ring with jangling knobs that bobbed up and down as she talked. She had a deep laugh, wore desert sandals, and her face made it quite impossible to say how old she was.

Wilma was not, in other words, anything like my preconception of what an Oxford tutor would be like. Even now, when I have met a good many other Oxford tutors and have even become one myself, I still find it impossible to fit Wilma into any known category of human being. She was not *like* anybody else. She had that rare quality of being utterly, completely, and naturally herself, of following her own furrow through life and commanding everybody's respect as she did so. That was why her death in 1989 was such a shock. I was not particularly close to her but she had always been *there*, an irreplaceable presence, belying any suggestion that women academics were not looked up to. She was always in great demand as a lecturer and tutor and I, one of a privileged band of her own students, took the respect in which she was held in the University completely for granted. (To use a term that has come into use since that time, she was, I suppose, a role model who was so good at being one that none of us realized that that was what she was.) It seemed quite natural to me that other students should consider that I had an unfair advantage in having Wilma as my tutor because I knew from my own experience just what memorable occasions her tutorials could be. They were full of eye-opening moments in which what you thought was true turned out not to be and in which arbitrariness and vagueness were exposed. They were not comfortable. You were made to feel annoyed with yourself for not having thought of something, but you remembered and thought better in the future. And you absorbed, even without consciously realizing it, her enthusiasm, her independence of thought, and her assumption that a lot of things were possible, like organizing expeditions into the desert or writing one of the standard textbooks on genetics. (Her first book, *Elementary genetics*, received the ultimate accolade of repeatedly being stolen from the library.)

In retrospect, I think I must have been a bit like one of Charles Darwin's children who, when being shown around the house of a friend asked 'And where does your father do his barnacles?' except that my unspoken question would have been 'And what is your tutor's next book going to be about?' At the time, Wilma's range of interests and abilities struck me as quite normal, just part of being a good biologist. But I see now that very few working scientists can claim to have written influential books on topics as diverse as genetics, biogeography, and medieval bestiaries, let alone gastronomy. (*Eating in eight languages* was published in 1968 and written while her leg was in plaster after a skiing accident.) Her many interests were not, however, put into separate compartments in her mind. As George Crowther, her husband and fellow explorer for 33 years put it, Wilma was always making unlikely connections and bringing different ideas together. She was an intellectual catalyst.

Wilma originally went up to Lady Margaret Hall, Oxford in 1937 to read Modern Languages but she soon switched to Zoology and graduated with a First-Class Honours degree. She then went to Girton College, Cambridge and did her doctoral thesis on genetics

('Phenocopies and their developmental significance in *Drosophila melanogaster*') between 1941 and 1943. She published *Elementary genetics* in 1951, after having worked as a statistician with Coastal Command at RAF Benson during the latter part of the war. Her knowledge of genetics and statistics provided the background for her later papers on rodent chromosomes, dispersal, and classification. Her first book on the distribution of animals, *Animal geography*, was published in 1962 and was subsequently reprinted five times. By this time, Wilma had become a lecturer at Somerville, a Fellow of Lady Margaret Hall, and a University lecturer at Oxford. She lectured on rodent biology, biogeography, desert ecology, evolution, elementary genetics, and the history of biology.

Through her interest in why different species have different distributions, Wilma became interested in one of the great early naturalists, Alfred Russel Wallace, whom she called the father of zoogeography. In 1964 she published a biography of Wallace, *Biologist philosopher*. In 1984, she and George went to the Malay Peninsula and crossed 'Wallace's Line' by boat, going from Bali across the Strait of Lombock. According to George, Wilma insisted on swimming, despite the sharks.

Once, on a walk through the galleries of the Vatican in Rome, Wilma noticed a sixteenth-century map of the world covered with pictures of animals. What struck her was not just that the pictures were accurately drawn, but that they were, remarkably, in the right places. Elephants were in Africa and there was none in South America. Someone, long before the term 'Zoogeography' had been thought of, had evidently known about the distribution of animals even though subsequently medieval map-makers came to be thought of as very ignorant on these matters. Jonathan Swift had mocked them for putting pictures of elephants simply to fill up the blank spaces on a map where there happened to be no towns. This view had been shared by most geographers until Wilma's book *Animals and maps: a study of the early cartographer's contributions to zoogeography* (1969) put the record straight. She showed that many of the early map-makers knew a great deal more about where different animals were to be found than their subsequent critics did themselves.

An invitation to give a lecture at the Mendel Museum in Brno led to a series of fruitful visits and regular publications in the journal *Mendelianum*. In 1975, Wilma published *Gregor Mendel and heredity*, which allowed her to combine her knowledge of the history of science with her long-standing knowledge of genetics. Meanwhile, she continued to go on expeditions to the world's deserts and to publish papers on her particular interest, gundis— Ctenodactylidae or 'comb toes'. She was also doing library research on bestiaries (an interest that followed on from *Animals and maps*) and continuing with a heavy teaching load. In 1982, while Wilma herself was in Western Australia observing rock wallabies, Fontana published her *Darwin* paperback, which Wilma always said should be exciting enough to read on a train.

Wilma had previously trapped rodents with George in the deserts of Medico and Arizona but she had always wanted to catch an animal that she had heard sat on a rock and groomed itself with combs that grew on its hind feet. There were other reports that it came out at night and sang to the moon. In 1966, she and George set off into the Sahara in search of these mysterious animals. Over several years, they brought back alive individuals of all four genera of gundis. The Ctenodactylids—with their satellite chromosomes that linked them to South America—turned out to be crucial in Wilma's research into rodent distribution. It is good to know that Wilma's chapter on the strange rodents of Africa and South America is being published in this volume under the editorship of René Lavocat, whom she greatly admired.

Wilma's chapter on the mammals of medieval bestiaries was published with Brunsdon Yapp's chapter on the birds in their joint book *The naming of the beasts* in 1991.

The news of Wilma's death was a sudden blow to all her many friends and colleagues. She had not told anyone that she was ill and had carried on as her usual vigorous ageless self. Six weeks before, she had been skiing in Courcheval, just as she had done for many years. She died on 14 March 1989.

Marian Stamp Dawkins
Somerville College, Oxford

(I would like to thank George Crowther for his help in writing this appreciation.)

WORKS BY WILMA GEORGE

BOOKS

George, W. (1951). *Elementary genetics* (2nd edn 1965). Macmillan, London.

George, W. (1962) *Animal geography.* Heinemann, London.

George, W. (1964). *Biologist philosopher: the life and writings of Alfred Russel Wallace.* Abelard–Schuman, London and New York.

George, W. (1969). *Animals and maps: a study of the early cartographer's contributions to zoogeography.* Secker and Warburg, London.

George, W. (1975). *Gregor Mendel and heredity.* Priory Press, London.

George, W. (1982). *Darwin.* Fontana, London.

George, W. and Yapp, B. (1991). *The naming of the beasts: natural history in the medieval bestiary.* Duckworth, London.

OTHER PUBLICATIONS

Biology

George, W. (1951). A statistic for an index of social conditions, circumstantial evidence for inbreeding. *British Journal of Sociology*, **2**, 255–9.

George, W. (1954). Cytology and evolution. *Entomologists Monthly Magazine*, **82**, 19–20.

George, W. (1955). Some animal reactions to variation of temperature. *Endeavour*, **12**, 5 pp.

George, W. (1970). Jird or gerbil? *Animals*, **12**, 475–7.

George, W. (1974). Desert ecology. *Open University course and Television.*

George, W. (1974). Notes on the ecology of gundis (family Ctenodactylidae). In *The biology of hystricomorph rodents* (ed. I. W. Rowlands and B. J. Weir), pp. 143–60. Academic Press, London.

George, W. (1977). Reproduction in female gundis (Ctenodactylidae, Rodentia). *Journal of Zoology, London*, **185**, 57–71.

George, W. (1978). World of animals and ecology of the living world. *Library of Modern Knowledge*, 328–30, 332–4, 384–414.

George, W. (1978). Combs, fur and coat care in the Ctenodactylidae (Rodentia). *Zeitschrift für Säugetierkunde*, **43**, 143–55.

George, W. (1979). The chromosomes of the hystricomorphous family Ctenodactylidae Rodentia: (? Sciuromorpha) and their bearing on the relationships of the four living genera. *Zoological Journal of the Linnean Society*, **65**, 261–80.

George, W. (1979). Conservatism of karyotype of two African mole rats (Bathyergidas). *Zeitschrift für Säugetierkunde*, **44**, 278–85.

George, W. (1980). The karyotypes of *Thryonomys gregorianus, Pedetes capensis* and *Hystrix cristata* and their relationships to other hystricomorphous and hystricognathous rodents. *Journal of the Linnean Society*, **68**, 361–72.

George, W. (1980). Caviomorph–hystricomorph karyotypes: ancestry or adaptation? *7th International Chromosome Conference*, Oxford, p. 73.

George, W. (1981). Species-typical calls in the Ctenodactylidae (Rodentia). *Animal Communication Symposium*, Cape Town and *Journal of Zoology*, **195**, 39–52.

George, W. (1981). Blood system patterns in rodents: contributions to an analysis of family relationships. *Zoological Journal of the Linnean Society*, **73**, 287–306.

George, W. (1981). Wallace and his line. In *Wallace's line and plate tectonics* (ed. T. C. Whitmore), pp. 3–8. Clarendon Press, Oxford.

George, W. (1981). The diet of *Petromus typicus* (Petromyidae, Rodentia) in Augrabies Falls National Park. *Koedoe*, **24**, 159–62.

George, W. (1982). *Ctenodactylus* (Ctenodactylidae, Rodentia): one species or two? *Mammalia*, **46**, 375–80.

George, W. (1984). Gundis on the 153. *153 Newsletter*, **21**, 5–7.

George, W. (1984). Gundis. In *The encyclopaedia of mammals* (ed. D. MacDonald), pp. 706–7.

George, W. (1985). *Wildlife at Lady Margaret Hall*, Lady Margaret Hall, Oxford.

George, W. (1985). Reproductive and chromosomal characters of ctenodactylids as a key to their evolutionary relationships. In *Evolutionary relationships among rodents* (ed. W. P. Luckett and J.-L. Hartenberger), pp. 453–74. Plenum Publishing, New York.

George, W. (1985). Cluster analysis of five species of Ctenodactylidae (Rodentia). *Mammalia*, **49**, 53–63.

George, W. (1986). The thermal niche: desert sand and desert rock. *Journal of Arid Environments*, **10**, 213–24.

George, W. (1986). Comparative anatomy and physiology of rodent stomachs. *Toxicology Forum, Geneva*, 247–57.

George, W. (1987). Water conservation in desert gundis (Ctenodactylidae). *African Small Mammal Newsletter*, 9.

George, W. (1987). Complex origins. In *Biogeographical evolution of the Malay archipelago* (ed. T. C. Whitmore), pp. 119–31. Clarendon Press, Oxford.

George, W. (1988). *Massoutiera mzabi* (Rodentia, Ctenodactylidae) in a climatological trap. *Mammalia*, **52**, 331–8.

George, W. and Crowther, G. (1981). Space-partitioning between two small mammals in a rocky desert. *Biological Journal of the Linnean Society*, **15**, 195–200.

George, W. and Weir, B. J. (1972). The chromosomes of some octodontids with special reference to *Octodontomys* (Rodentia, Hystricomorpha). *Chromosome*, **37**, 53–62.

George, W. and Weir, B. J. (1972). Record chromosome number in a mammal? *Nature New Biology*, **236**, 206–6.

George, W. and Weir, B. J. (1973). A note on the karyotype of *Proechimys quairae* (Rodentia, Hystricomorpha). *Mammalia*, **37**, 330–2.

George, W. and Weir, B. J. (1974). Hystricomorph chromosomes. In *The biology of hystricomorph rodents* (ed. I. W. Rowlands and B. J. Weir), pp. 79–108. Academic Press, London.

George, W., Weir, B. J., and Bedford, J. (1972). Chromosome studies in some members of the family Caviidae (Mammalia, Rodentia). *Journal of Zoology, London*, **168**, 81–9.

History of science

George, W. (1964). The use of biogeography in authenticating early discoveries. *Geographical Journal*, **130**, 315–17.

George, W. (1964). An early description of an Australasian mammal. *Nature, London*, **202**, 1130–1.

George, W. (1965). An early European description of an Australian mammal. *Nature, London*, **205**, 517–18.

George, W. (1967). Mendel and the classification of mammals. *Folia Mendeliana*, **2**, 23–8.

George, W. (1968). The yale. *Journal of the Warburg and Courtauld Institute*, **31**, 423–38.

George, W. (1970). Vitus Bering. *Dictionary of Scientific Biography*, **2**, 14–15.

George, W. (1970). Jean-Baptiste Bory de Saint-Vincent. *Dictionary of Scientific Biography*, **2**, 320–1.

George, W. (1970). Louis Bougainville. *Dictionary of Scientific Biography*, **2**, 342–3.

George, W. (1971). Francis Darwin. *Dictionary of Scientific Biography*, **3**, 581–2.

George, W. (1971). The reaction of A. R. Wallace to the work of Gregor Mendel. *Proceedings of the Gregor Mendel Symposium*, 173–7.

George, W. (1978). The fauna of the printed map. *The Map Collector*, **5**, 2–8.

George, W. (1980). Sources and background to the discovery of new animals in the sixteenth and seventeenth centuries. *History of Science*, **18**, 79–104.

George, W. (1980). The gentle trader: collecting in Amazonia and the Malay Archipelago 1848–1862. *Journal of the Society for the Bibliography of Natural History*, **9**, 503–14.

George, W. (1981). The bestiary: a handbook to the local fauna. *Archives of Natural History*, **10**, 187–203.

George, W. (1982). Gregor Mendel and Andreas von Ettingshausen. *Folia Mendeliana*, **17**, 213–16.

George, W. (1982). The Darwinians and the role of women. In *Social history of the bio-medical sciences* (ed. M. Piatelli-Palmarini), Milan.

George, W. (1982). Wallace and the birds: changing classification 1848–1874. *History in the Service of Systematics colloqium*.

George, W. (1982). The Mendel enigma, the farmer's son: the key to Mendel's motivation. *Archives internationales d'Histoire des Sciences*, **32**, 177–83.

George, W. (1983). The making of Mendel. In *Gregor Mendel and the foundation of genetics* (ed. V. Orel and A. Matalova), pp. 270–87.

George, W. (1984). Mendel, *fils de paysan*. In *Le cas* (ed. J.-R. Armogathe), pp. 17–25. Université de Paris-Sud.

George, W. (1985). The living world of the bestiary. *Archives of Natural History*, **12**, 161–4.

George, W. (1986). Alive or dead: zoological collections in the seventeenth century. In *The origins of museums* (ed. O. Impey and A. MacGregor), pp. 179–87. Clarendon Press, Oxford.

George, W. (1988). The zoology of Ulisse Aldrovandi. In *Tribute to Aldrovandi* (ed. T. Tomasi and G. Olmi). Coeckelberghs Verlag, Lucerne.

George, W. (1988). Thomas Harriot and the fauna of North America. *The Durham Thomas Harriot Seminar Occasiona*, paper 6, 1–32.

George, W. (1988). A. R. Wallace: Evolution and the distribution of animals. *Biology History*, **2**, 5–6.

1 THE HISTORY OF THE PROBLEM
Albert E. Wood

The relationships of South American organisms involve several distinct problems regarding organisms emplaced in South America at quite different times. There is no a priori reason that they should have involved similar geographic relationships of faunas or similar types of migration routes.

South American organisms may be divided into several categories (termed faunal strata by Simpson 1950). The explanation for the origin of one level is independent of that for the others.

1. There are organisms presumably descendants of ones that had had a pan-Gondwanaland distribution before the split between Africa and South America (*c.* 100–120 my ago, Duncan and Hargraves 1984), including

Nothofagus and other early flowering plants . . .; chironomid midges and parastacoid crayfishes . . .; a remarkable majority of its freshwater fishes including characids, cichlids, and both loricariid and bunocephalid catfishes . . .; such amphibians as the hylids, pipids, leptodactylids, and caecilians; chelid and pelomedusid turtles; amphisbaenid and gekkonid lizards; . . . and such avian groups as ratites, penguins and several suboscine groups . . . In each instance where the time-range is well known, Gondwana groups had evolved by late Jurassic or early Cretaceous time (Webb 1978).

This group also includes lung-fish. Cretaceous titanosuchid dinosaurs are known only in South America and Africa (Pascual *et al.*,1985) except for the late Cretaceous *Alamosaurus* of New Mexico, probably an immigrant stock from South America.

2. Some animals known from the late Cretaceous of South America show marked relationships to ones from North America (*Kritosaurus*, an otherwise North American hadrosaur genus from the Argentine; didelphoid marsupials; a possible condylarth; (Bonaparte 1984; Estes and

Baéz 1985; Pascual *et al.* 1985; Van Valen 1988); others represent possible endemic derivatives of unknown ultimate origin (Scillato-Yané and Pascual 1984). Marsupials may have originated in South America, North America, Antarctica, or Australia, all of which must have been sufficiently interconnected at this time to have permitted intermigration. The ungulates are most likely of North American origin. There are possible edentates in the Cretaceous faunas, which were probably an endemic South American group, of uncertain relationships (Simpson 1978). Other Cretaceous mammals from South America are so poorly known that their relationships can only be guessed at (Pascual *et al.* 1985; Van Valen 1988).

3. Ceboid primates and caviomorph rodents appear (the latter already highly diversified) in the late Oligocene Deseadan of Patagonia and Bolivia. Both groups were unquestionably invaders by island-hopping, either from Africa (Hoffstetter and Lavocat 1970; Hoffstetter 1971, 1972; Lavocat 1971, 1974) or Middle America (Wood 1950, 1980, 1985; Patterson and Wood 1982). Also appearing at the same time is a frog, possibly with African relationships (Estes and Baéz 1985).

4. What Webb (1978) has termed 'The Great American Interchange' included major faunal interchanges between North and South America that began just before the completion of the Panamanian land bridge, but which reached a climax after the land bridge became continuous, with many types of mammals migrating both north from South America, but more especially south from North America to dominate the Pleistocene and Recent South American faunas.

5. Hershkovitz (1972) considered the South American cricetine rodents as representing a distinct stratum. Most authors believe that the

most likely explanation for their distribution is that they had differentiated extensively in Middle America before crossing by way of Panama as an early part of the Great American Interchange. Although Hershkovitz doubted that this would have been possible, a significant number of members of the group are known from the northern edge of Middle America, in the southern United States (Baskin 1978). The main mass of Middle America is *terra incognita* as far as Tertiary mammals are concerned.

These, then, are the types of distribution that authors have attempted to explain. It would not be possible to include all the explanations that have been proposed, but the following includes the most important.

The first item in Wilma George's notes on this introduction was 'From Acosta (1589) to the present day'. José de Acosta was a Spanish Jesuit who spent 20 years (1565–1585) in the Spanish Indies. On returning to Spain, he published the information he had gathered in the New World as the *Natural history of the Indies*, in 1588, and the *Conversion of the Indians*, in 1589 in Salamanca (both titles translated). These were combined in a single publication as the *Historia natural y moral de las Indias*, published at Seville in 1590. Translations were published, from 1597 to 1607, in French, German, Dutch, Latin, English, and Italian. The English translation, published in 1604 by the Hakluyt Society, is the basis for what follows.

De Acosta was a shrewd observer. He noted (especially while he was in Peru) that there were many animals that differed from those of the Old World; that there were others that were similar; and others that had been imported by the Spaniards. He saw no way by which Old World animals (from Europe, Asia, or Africa) would have been able to cross the Atlantic by themselves, and for some of them (for example wolves and foxes) he did not believe that humans would have brought them in their boats. So he felt that there must be land connections (thus far undiscovered) between the Old and New Worlds,

either in the Arctic or the Antarctic regions. He was worried that some animals were peculiar to the New World, and unlike anything known in the Old World. He could imagine that they came overland as had the others, but how did it happen that they left no trace of their passage in the form of stragglers? In addition, they would have had to spread very rapidly to have reached South America following the Noachian deluge. He considered the possibility that, after the Flood, they represented a new creation by the Deity, but felt that this would seem to be contrary to the account in the *Book of Genesis* and to Christian teaching. He ultimately left the matter unresolved.

In the *Histoire naturelle* (1749–1804), Buffon suggested (1749) that there had perhaps been land connections between Africa and South America, because of their faunal similarities. However, he noted that there were also obvious dissimilarities in their faunas, and was uncertain as to how to explain matters.

Cuvier recognized (1812) that there were close similarities between South American mastodonts, brought back by Humboldt, and European ones. To Cuvier, this was merely an interesting observation, since he believed firmly in the fixity of species, and explained the distinct nature of fossils as being representatives of previous creations that had been removed by catastrophic extinctions (Simpson 1984).

Forbes (1846) proposed land bridges to explain similarities between areas now separated. For example, he postulated land bridges between Britain and various parts of Europe, to account for assorted items of the biota. He believed, more generally, that all islands had had, at one time or another, land connections to the nearest continent. However, he did not discuss the problems of South American–African relationships. Forbes may be considered to be the founder of the school of land bridges.

In the *Origin of species* (1859), Darwin stressed the fundamental differences ('utterly dissimilar') between the faunas and floras of Africa, South America, and Australia, even though they were in the same latitudes and possessed similar environmental conditions.

Likewise, Wallace (1876) was sure that there was no evidence for a former land connection between South America and Africa, and that such similarities as there were between the organisms of the two areas represented corresponding adaptations in similar climates. He followed Darwin in dividing islands into oceanic, that had evolved their biota in isolation, and continental ones, whose life was essentially like that of the adjacent continents, from whence it had come.

After the acceptance of Darwin's theory of evolution, the explanation for biotic similarities between different parts of the world became a problem that obviously required explanation. Equally important was finding the explanation for faunal and floral differences.

Gradually, it was realized that some present-day barriers (especially, for example, the Bering Straits) had not been present at all times in the past. But it was accepted that major oceans had been barriers to the migration of terrestrial organisms. Geological studies of the earth's structure and history (and especially the general acceptance of the theory of isostasy, according to which high-standing areas of the earth's crust were underlain by lighter rocks than were low-standing areas) indicated that there had been long-lasting differences between the continents and the ocean basins.

In Chapter II of *The geographical distribution of animals* (1876), Wallace explored all the ways in which organisms might have been distributed, and particularly all possible means at their disposal to cross extensive water barriers. For many smaller organisms, both plant and animal, wind transport was possible. This was also true of spores or larval stages. Drifting debris could carry seeds or living animals over considerable distances. Birds could transport seeds or spores of plants or eggs and larvae of animals, in mud on their feet, in their feathers, as internal parasites, or as still viable reproductive structures in their digestive systems. Small barriers could be crossed, especially by mammals, by swimming.

These views have been supported by the studies of the repopulation of Krakatoa during the early part of the 20th century, following the complete destruction of life on the island by the volcanic eruption of 1883, which demonstrated how rapidly these processes could operate over short distances.

In the latter part of the 19th and the early part of the 20th centuries, land bridges became a very popular way to explain any cases of biological similarity between disjunct areas, even to include cases of single species that seemed to be out of place. Von Ihering (1907) felt that it was impossible to explain the fauna and flora of St Helena without a land bridge from Africa. Joleaud 1919*a*) postulated a late Miocene land bridge from Spain and North Africa to Florida and the Antilles, to account for isolated horse teeth found in North Carolina and Florida that Gidley had referred to the Old World horse genus *Hipparion*. Having 'built' this land bridge, Joleaud expanded its operations so that most mid- to late-Tertiary New World to Old World intermigrations took place over it, and so that it permitted migration both ways between Europe and both North and South America, but not between the last two (Joleaud 1919*b*).

W. D. Matthew (1915), in accordance with the then-general viewpoint, was a firm believer in the fixity of continents. He believed that the main centre of mammalian evolution had been Holarctica, and that the southern continents had been populated by migrations from the north, over the routes that are currently usable. In a majority of cases, these conclusions were clearly supported by the presence of lineages in Holarctica before they were found in the southern hemisphere.

Wegener (esp. 1924) indicated that the lock-and-key resemblance of the coastlines of Africa and South America was an indication that they had formerly been united and had split apart. He continued the former relationship across the North Atlantic, between Europe and North America, as well. Although there were a number of important adherents to this theory, particularly in Europe, the very general objection was that there was no known mechanism for the migration of continents, and that it was contrary to all that was then known of the structure of the earth's crust.

In a number of papers (1940, 1943, and others), Simpson, following Matthew's general principles, discussed intercontinental migrations, especially of land mammals. He argued (1940) against the indiscriminant creation of land bridges. He pointed out that connections between two areas would vary from corridors, permitting a complete interchange of faunas and floras (as between western Europe and China at the present time); to narrow connections, termed land bridges, permitting migration by all organisms whose geographic ranges reached the abutments of the bridge, resulting in migration not of a single form but of a balanced fauna and flora; to a discontinuous connection, normally a chain of islands, which would permit migration by only a scattering of terrestrial forms. He pointed out (1943) that an absence of broad similarities between two areas could be considered as proof that there was no continuous land connection between them.

Over the years, a number of authors have identified similarities between South American and African forms that they believe represented evolutionary relationships, without specifying the migration routes that were involved. Florentino Ameghino (1906), describing the Tertiary mammals of Argentina, reached the conclusion (strongly aided by a misidentification of their geologic ages) that these were ancestral to the morphologically similar animals of the rest of the world, which would have required some type of extensive migration routes, but what these were he did not specify.

Hershkovitz (1972), not believing that the North American hesperomyine cricetids could have given rise to the large number of genera that flooded South America at about the time of the establishment of the Panamanian land bridge, preferred to believe that they represented an early Tertiary invasion (by rafting) from Africa, in spite of the fact that no known potential ancestors were known in that continent. He previously (1969) had accepted a North or Middle American origin for the group.

Croizat (1979) proposed that at

some unspecified time earlier than the beginning of the Tertiary, certain very primitive 'proto-rodents' originated in a then unitarian continent of Gondwana. When this continent split into a later South America and Africa, respectively, these 'proto-rodents', already on the way to evolving as taxonomically definable 'hystricomorph rodents' turned 'American' and 'African' by right of the new continental geography.

Ellerman (1940), in his monograph on the rodents of the world, referred many South American caviomorphs to separate African subfamilies. When I queried such a situation because it would have required multiple invasions from Africa, he replied (*in lit.*) that he was only interested in the morphology of the animals, and not in their possible migration routes.

Subsequent to World War II, the development of capabilities for drilling, seismic exploration, and extensive dredging in the deep oceans resulted in a greatly improved knowledge of the history of these oceans. This led to the currently accepted theory of plate tectonics—that there are major rifting zones in the central part of oceans (along which lava rises from the mantle lying at shallow depths beneath the ocean bottom). The rising lavas push the rifting ocean crust apart at significant rates, carrying the continents along ahead of the ocean crust. At the same time, the leading edges of the continents collide with other oceanic plates, whose denser rocks are subducted beneath the continental margins, elevating the overriding continents into mountain chains.

Plate tectonics is accepted as having resulted in the separation of Africa and South America, starting near the beginning of the Cretaceous, at about 120 my ago. North America had begun to split from Africa (and, to a lesser extent, from South America) in the Jurassic. A volcanic island arc arose, between southern North America (Middle America) and South America, along the line of the present Panama land bridge, by the early Cretaceous (Duncan and Hargraves 1984). This arc permitted the migration of terrestrial organisms between North and South America through much of the Cretaceous. A section of Pacific ocean floor, spreading from a hot spot beneath the Galapagos, began to move eastward as the Caribbean Plate, pushing the island arc eastward in the late Cretaceous (80 my ago). This movement of the Caribbean Plate continued throughout the Cenozoic, the island arc gradually being pushed farther, and its islands faulted and sliced and elongated, with the major part becoming the Greater Antilles and the southern part being rotated into South America. During this time, the ends of the arc maintained reasonably close contact with both Middle and South America. In this process, the Caribbean Plate moved a total of over 1400 km eastward with respect to both North and South America (Burke *et al.* 1984). As the Caribbean Plate moved between North and South America, a second volcanic island arc began to form behind it, being in place along the line of the present Panama land bridge by the middle of the Cenozoic (Duncan and Hargraves 1984).

During its western migration, the leading edge of South America accreted a considerable number of allochtbonous blocks from sources to the west or north (Case *et al.* 1984), which might have provided the means of transportation of organisms that McKenna (1973) has termed 'Noah's arks or beached Viking funeral ships'.

In recent years, two schools have vigorously presented opposing views of the origins of the ceboid primates and caviomorph rodents of South America, groups not known anywhere before the late Oligocene (MacFadden 1985) Deseadan. One, represented by Hoffstetter and Lavocat (Hoffstetter and Lavocat 1970; Hoffstetter 1971, 1972; Lavocat 1971, 1974) derives them from unknown African forms, similar to, but more primitive than, members of the well-known early Oligocene fauna of the Fayum, Egypt, also represented by fragmentary material from the North African Eocene. The other school, represented by Wood and Patterson (Wood 1950,

1980, 1985; Patterson and Wood 1982), derives them from forms that they believe were developing (but thus far only known from a scattering of peripheral members) in Middle America. The Hoffstetter–Lavocat proposal involves crossings (at least two) of the South Atlantic, then a distance of 3500 km (Webb 1978). Tarling (1980) proposed that those crossing took place by island-hopping, using islands presumed to have been present (but not yet identified as islands by drilling) on the Céara and Sierra Rises. The Wood–Patterson hypothesis likewise involves at least two crossings, also by island-hopping, either through the Greater Antilles or along the developing pre-Panama island arc. This would presumably have involved a water-gap of not more than 500 km, significant portions of which were interrupted by islands (Webb 1978).

The final solution to the source of the ceboids and caviomorphs will depend on the discovery of pre-Deseadan fossil members of these groups in South America, as well as Middle American discoveries to show what, if any, possible ancestors lived there in the Eocene or early Oligocene.

ACKNOWLEDGEMENTS

Wilma George had obviously thought extensively on the historical background, but, given her interest in the history of biology (see George 1964), but the material that was made available to me as being her data for this introduction consisted basically of a few names, dates, and key phrases. I have attempted to follow her outline, although I have been handicapped by being at a considerable distance from any research library. I could not have written this chapter without the assistance of the Cape May County Public Library, and particularly Mrs Pat Fagan, Head of Interlibrary Loans, who obtained all the books that I requested.

REFERENCES

Acosta, José de. (1590). *Historia natural y moral de las Indias*. Seville. (English translation: *Natural and moral history of the Indies*, published by the Hackluyt Society, 1604.)

Ameghino, Florentino. (1906). Les formations Sédimentaires du Cretacé Supérieur et du Tertiaire de Patagonie avec un Parallèle entre leurs Faunes Mammalogiques et celles de l'Ancien Continent. *Anales del Museo Nacional de historia natural de Buenos Aires*, (3) **15**, 1–568.

Baskin, J. (1978). *Bensonomys*, *Calomys*, and the origin of the phyllotine group of Neotropical cricetines (Rodentia: Cricetidae). *Journal of Mammalogy*, **59**, 125–35.

Bonaparte, J. F. (1984). Nuevas pruebas de la conexión fisica entre Sudamérica y Norteamérica en el Cretácico tardio (Campaniano). *Actas III Congreso Argentino de Paleontologia y Bioestratigraphia*, 141–8.

Buffon, G. L. L. de. (1749–1804). *Histoire naturelle*. Paris, 15 vols, Suppl., 7 vols.

Burke, K., Cooper, C., Dewey, J. F., Mann, P., and Pindell, J. L. (1984). Caribbean tectonics and relative plate movements. In *The Caribbean–South American Plate boundary and regional tectonics* (ed. W. E. Bonini, R. B. Hargraves, and R. Shagam), *Geological Society of America*, Memoir **162**, 31–63.

Case, J. E., Holcombe, T. L., and Martin, R. G. (1984). Map of geologic provinces in the Caribbean region. In *The Caribbean–South American Plate boundary and regional tectonics* (ed. W. E. Bonini, R. B. Hargraves, and R. Shagam), *Geological Society of America*, Memoir **162**, 1–30.

Croizat, L. (1979). Review of *Biogeographie: Fauna und Flora der Erde und ihre geschichtliche Entwicklung*, by P. Bănărescu and N. Boşcaiu (1976). Gustav Fischer, Jena. *Systematic Zoology*, **28**, 250–2.

Cuvier, G. L. C. F. D. (1812). *Recherches sur les ossements fossiles, où l'on rétablit les caractères de plusieurs animaux dont les revolutions du globe ont détruit les espèces* (1st edn), 4 vols., Paris.

Darwin, Charles. (1859). *The origin of species by means of natural selection, or the preservation of favoured races in the struggle for existence*. London, Murray.

Duncan, R. A. and Hargraves, R. B. (1984). Plate tectonic evolution of the Caribbean region in the mantle reference frame. In *The Caribbean–South American Plate boundary and regional tectonics* (ed. W. E. Bonini, R. S. Hargraves, and R. Shagam), *Geological Society of America*, Memoir **162**, 81–93.

Ellerman, J. R. (1940). The families and genera of living rodents. *British Museum* (*Natural History*), **1**, 1–689. London.

Estes, R. and Baéz, A. (1985). Herpetofaunas of North and South America during the Late Cretaceous and Cenozoic: Evidence for Interchange? In *The Great American Biotic Interchange* (ed. F. G. Stehli and S. D. Webb), pp. 140–97. Plenum Press, New York, London.

Forbes, E. (1846). On the connection between the distribution of the existing fauna and flora of the British Isles, and the geological changes which have affected their area, especially during the epoch of the Northern Drift. *Distr. Mem. Geol. Surv. Gt. Britain*, [Great Britain. Geological Survey (District memoirs)] **1**, 336–432.

George, Wilma. (1964). *Biological philosopher: a study of the life and writings of Alfred Russel Wallace*. Abelard–Schuman, London, Toronto, New York.

Hershkovitz, R. (1969). The evolution of mammals on southern continents. VI. The Recent mammals of the Neotropical Region: a zoological and ecological review. *Quarterly Review of Biology*, **44**, 1–70.

Hershkovitz, P. (1972). The Recent mammals of the Neotropical Region: a zoogeographic and ecological review. In *Evolution, mammals and southern continents* (ed. A. Keast, F. C. Erk, and B. Glass), pp. 311–431. State University of New York Press, Albany, NY.

Hoffstetter, R. (1971). Le peuplement mammalien de l'Amérique du Sud. Rôle des continents austraux comme centre d'origine, de diversification et de dispersion pour certains groupes mammaliens. *Anals da l'Academia Brasileira de Ciencias*, **43** (Suppl.), 125–44.

Hoffstetter, R. (1972). Origine et dispersion des rongeurs hystricognathes. *Comptes Rendus de l'Académie des Sciences Paris* (D), **274**, 2867–70.

Hofstetter, R. and Lavocat, R. (1970). Découverte dans le Déséadien de Bolivie de genres pentalophodontes appuyant les affinités africaines des Rongeurs Caviomorphes. *Comptes Rendus de l'Académie des Sciences Paris* (D), **271**, 172–5.

Ihering, H. von. (1907). *Archhellenis* und *Archinotis*. 350 pp. Leipzig.

Joleaud, L. (1919a). Relations entre les migrations du genre *Hipparion* et les connexions continentales de l'Europe, de l'Afrique et de l'Amérique au Miocène supérieur. *Comptes Rendus de l'Académie des Sciences Paris*, **168**, 177–9.

Joleaud, L. (1919*b*). Sur les migrations des genres *Hystrix*, *Lepus*, *Anchitherium* et *Mastodon* à l'époque néogène. *Comptes Rendus de l'Académie des Sciences Paris*, **168**, 412–14.

Lavocat, R. (1971). Affinités systématiques des cavio-morphes et des phiomorphes et origine Africaine des caviomorphes. *Anals de l'Academia Brasileira de Ciencias*, **43** (Suppl.), 515–22.

Lavocat, R. (1974). The interrelationships between the African and South American rodents and their bearing on the problem of the origin of South American monkeys. *Journal of Human Evolution*, **3**, 323–6.

MacFadden, B. J. (1985). Drifting continents, mammals and time scales; current developments in South America. *Journal of Vertebrate Paleontology*, **5**, 169–74.

McKenna, M. C. (1973). Sweepstakes, filters, corridors, Noah's arks, and beached Viking funeral ships in palaeogeography. In *Implications of continental drift to the earth sciences* (ed. D. H. Tarling and S. R. Runcorn), Vol. 1, pp. 295–308. Academic Press, London.

Matthew, W. D. (1915). Climate and evolution. *Annals of the New York Acadademy of Sciences*, **24**, 171–318.

Pascual, R., Vucetich, M. G., Scillato-Yané, G. J., and Bond, M. (1985). Main pathways of mammalian diversification in South America. In *The great American biotic interchange* (ed. F. G. Stehli and S. D. Webb), pp. 219–47. Plenum Press, New York and London.

Patterson, B. and Wood, A. E. (1982). Rodents from the Deseadan Oligocene of Bolivia and the relation-ships of the Caviomorpha. *Bulletin of the Museum of Comparative Zoology*, **149**, 371–543.

Scillato-Yané, G. J. and Pascual, R. (1984). Un peculiar Xenarthra del Paleocene medio de Patagonia (Argentina). Su importancia en el sistematica de las Paratheria. *Ameghiniana*, **21**, 2–4.

Simpson, G. G. (1940). Mammals and land bridges. *Journal of the Washington Academy of Sciences*, **30**, 137–63.

Simpson, G. G. (1943). Mammals and the nature of continents. *American Journal of Sciences*, **241**, 1–31.

Simpson, G. G. (1950). History of the fauna of Latin America. *American Scientist*, **38**, 361–89.

Simpson, G. G. (1978). Early mammals in South America: fact, controversy, and mystery. *Proceedings of the American Philosophical Society*, **122**, 318–28.

Simpson, G. G. (1984). *Discoverers of the Lost World*. Yale Univ. Press, New Haven and London.

Tarling, D. H. (1980). The geologic evolution of South America with special reference to the last 200 million years. In *Evolutionary biology of the New World monkeys and continental drift* (ed. R. L. Ciochon and A. B. Chiarelli), pp. 1–41. Plenum Press, New York and London.

Van Valen, L. (1988). Faunas of a southern world. *Nature*, **333**, 113.

Wallace, A. R. (1876). *The geographical distribution of animals with a study of the relations of living and extinct faunas as elucidating the past changes of the earth's surface*. Macmillan, London.

Webb, S. D. (1978). A history of savanna vertebrates in the New World. Part II: South America and the great interchange. *Annual Review of Ecological Systems*, **9**, 393–426.

Wegener, A. (1924). *The origin of continents and oceans*. Methuen and Co., London.

Wood, A. E. (1950). Porcupines, paleogeography and parallelism. *Evolution*, **4**, 87–98.

Wood, A. E. (1980). The origin of the caviomorph rodents from a source in Middle America: a clue to the area of origin of the Platyrrhine primates. In *Evolutionary biology of the New World monkeys and continental drift* (ed. R. L. Ciochon and A. B. Chiarelli), pp. 79–91. Plenum Press, New York and London.

Wood, A. E. (1985). The relationships, origin and dispersal of the hystricognathous rodents. In *Evolutionary relationships among rodents. A multidisci-plinary analysis* (ed. W. P. Luckett and J.-L. Hartenberger), NATO Advanced Science Institutes (A), **92**, pp. 475–513. Plenum Press, N.Y.

2 THE PALAEOGEOGRAPHY OF THE OPENING SOUTH ATLANTIC

Judith Totman Parrish

INTRODUCTION

The history of South Atlantic palaeogeography begins with the supercontinent, Pangaea. The assembly of Pangaea began with the middle Palaeozoic collision of north-western Europe with North America on the west and the Siberian platform on the east to form the continent called Laurussia. Shortly thereafter (geologically), Gondwana collided with Laurussia. The final assembly of Pangaea occurred in the Early Triassic with the collision between North and South China that raised the Chinling Range (J. M. Parrish *et al.* 1986). No sooner had Pangaea come together than it began breaking up. Continental deposits associated with the initial rift valleys in eastern North America are Triassic in age. The Central Atlantic opened first between north-western Africa and south-eastern North America. Because South America and Africa were still one plate, the opening of the Central Atlantic caused relative motion between North and South America; the Gulf of Mexico and eventually the Caribbean formed in the intervening space. The opening of the South Atlantic began in the Early Cretaceous. The final plate separation in the circum-Atlantic region was between North America and north-western Europe. The central focus of this book is the history of connections between Africa and South America, and so the history of the South Atlantic forms the principal focus of this chapter. However, connections between North and South America and Africa and Europe (and thus North America) are also discussed.

SOUTH ATLANTIC

Sea-floor spreading history

The rate of subsidence of new oceanic crust is proportional to the square root of its age (Sclater *et al.* 1977, and references therein). In the normal oceanic crust of a mature ocean basin, the ridge crest lies approximately 2500 m below sea-level, with the depth increasing away from the crest in a curve related to the subsidence rate. This simple relationship is complicated by the additional sub-sidence caused by sediment loading, by the build-ing of volcanic piles on top of the normal crust, by tectonic events not related to spreading, and by the presence of anomalously shallow areas of various types.

The subsidence curve is used as a basis for reconstructing the palaeodepths of particular sites in the oceans, and some of these will be discussed later. Sclater *et al.* (1977) constructed maps outlining the sea-floor spreading history of the South Atlantic using the subsidence curve. For the purpose of this discussion, the key features of their reconstructions include not only the bathymetric evolution of the major deep basins in the South Atlantic, but also the evolution of the Gulf of Guinea, the growth and subsidence of Walvis Ridge and the Rio Grande Rise, and the opening of the southernmost South Atlantic east of the Falkland Plateau. The subsidence curve is most reliable for crust 80 million years old and younger (older crust tends to be around 500 m shallower than predicted by the curve). Sclater *et al.* (1977) maintained that their palaeodepth predictions from normal young oceanic crust are accurate to within 300 m. The reader should bear in mind that the work of Sclater *et al.* (1977) applies to the formation and subsidence of the oceanic crust; continental and oceanic geology will be discussed in greater detail in following sections of this chapter.

Prior to the Albian Age, the South Atlantic was still closed but the Central Atlantic was already open, having begun opening at least by the Late Jurassic (Pitman *et al.* 1974), and probably earlier along a spreading centre west of the Blake Spur that subsequently was aborted (Vogt 1973; Klitgord and Grow 1980; Pindell *et al.* 1988). The formation of the Central Atlantic included the separation of Florida, originally part of Gondwana, from the notch between north-eastern South America and north-western Africa (Pindell and Dewey 1982). The oldest magnetic anomaly in the South Atlantic is about 125 million years, or Hauterivian age on the Harland *et al.* (1982) time-scale (Sclater *et al.* 1977). Sclater *et al.*'s (1977) first reconstruction after the initiation of South Atlantic rifting is for the Albian Age, about 110 million years before present (mybp). By the Albian, the Brazil–Angola and Argentine–Cape Basins had already subsided below 3000 m. Sclater *et al.*'s (1977) reconstruction shows a gap approximately 50 km wide between the 200 m depth contours off northern Ceará State in Brazil and off Ghana on the African side. They included a 2000 m depth contour at that point (see also Thiede 1979).

Agreement with Sclater *et al.* (1977) on the timing of the opening of the equatorial Atlantic during Aptian–Albian is widespread, although the details of other workers' interpretations vary. Principally, inclusion of the Amazon Delta sediment wedge in the continental outlines used for refitting the margins is now rejected (Rabinowitz and LaBrecque 1979; Pindell and Dewey 1982; Pindell *et al.* 1988; and others). Rather, extension of the Benue Trough during rifting is now recognized to have taken place. Rabinowitz and LaBrecque (1979) had the equatorial Atlantic opening first along the northern Brazilian margin, forming Africa into its present shape, and then along the rest of the South Atlantic rift. In contrast, Pindell and Dewey (1982) and Pindell *et al.* (1988) preferred the interpretation that the South

Atlantic opened first, creating Africa's present configuration, followed by the equatorial Atlantic. Pindell and Dewey's (1982) and Pindell *et al.*'s (1988) model explains the fan-like opening of the South Atlantic, which opened first in the southern part. Rabinowitz and LaBrecque (1979) and Pindell and Dewey (1982) placed the opening of the Gulf of Guinea at about 110 mybp; Pindell *et al.* (1988) cited 119 mybp for initial divergence. However, the oldest magnetic anomaly in the region is early Campanian. Between the Aptian and early Campanian, magnetic data are lacking as a result of the Cretaceous Quiet Period and possibly also because of deformation associated with transform-margin motion (Pindell *et al.* 1988).

Other features of Sclater *et al.*'s (1977) reconstruction for the Albian include a wide corridor above 200 m palaeodepth between South America and Africa in the central South Atlantic. This shallow area was composed of what are now the landward extremities of the Walvis Ridge and the Rio Grande Rise. In the southern South Atlantic, Sclater *et al.* (1977) did not extend their reconstruction beyond the middle of the Falkland Plateau. Nevertheless, they showed water depths up to 2000 m between South Africa and the easternmost part of the Plateau.

By the Cenomanian, about 95 mybp, water depths between northern Brazil and Ghana had increased to as much as 4000 m in some basins. Although deep-water flow was likely to have been restricted by topographically high fracture-zone ridges (not figured by Sclater *et al.* 1977, but discussed by them), no evidence exists that these ridges stood above sea-level. The Walvis Ridge–Rio Grande Rise connection was still very shallow, mostly less than 200 m, but with a slightly deeper gap just east of the São Paulo Plateau. The water around the Falkland Plateau had deepened to 3000–4000 m and, like the fracture zones to the north, the Falkland Fracture Zone probably was not emergent, although it could have been a barrier to deep-water circulation.

Sclater *et al.*'s (1977) reconstruction for the

Campanian Stage (80 mybp) shows water depths of 2000 m for parts of the Walvis Ridge–Rio Grande Rise complex and does not include 200 m isobaths. The conclusion, then, is that if South Atlantic bathymetric evolution followed the theoretical subsidence curve, all connection between Africa and South America was severed sometime between 95 and 80 mybp, during the early Late Cretaceous. However, Sclater *et al.*'s (1977) palaeodepth predictions did not take sea-level changes into account and they allowed a possible error in palaeodepth reconstruction of as much as 300 m. A 300 m error and/or a significant change in sea-level would be particularly critical for determining the submergence history of the Walvis Ridge–Rio Grande Rise and Sierra Leone Rise–Ceará Rise–Gulf of Guinea regions. In addition, Austin and Uchupi (1982) suggested

that the earliest magnetic anomaly in the South Atlantic was not preserved in true oceanic crust, and that the formation of true oceanic crust post-dated the first anomaly by 4–9 million years; this would move the final severance of the connection to as late as the Maastrichtian. Additional clues must be garnered from rocks that have been recovered from the South Atlantic, subsidence curves reconstructed for specific South Atlantic sites, and adjacent continental geology.

Deep-sea drilling and dredge sites

Selected South Atlantic core sites and the age of the oldest rocks drilled at each one are listed in Table 2.1; their positions are indicated on Figure 2.1. The sites of greatest interest are those from the Ceará and Sierra Leone Rises and the

Table 2.1 List of DSDP sites discussed in text and the oldest sediment drilled at each

Location	Site	Oldest sediments penetrated
Ceará Rise	354	Maastrichtian, deep-water sediments
Sierra Leone Rise	13A	early Early Cretaceous, deep-water shale and chert
Rio Grande Rise	21	Campanian, shell debris containing red algae
	22	Eocene, deep-water sediments
	357	Middle Eocene, shallow-water fossils and subaerially weathered volcanic rocks
	516	Coniacian, shallow-water sediments, water depth as little as 20 m
São Paulo Plateau	356	Late Albian, fossils indicating 500–1500 m palaeodepth
Walvis Ridge	359	Eocene, subaerially extruded volcanics
	362	Early Eocene, deep-water sediments
	363	Late Aptian, calcareous sand and algae
	525A	Campanian pelagic sediments
	526	Late Palaeocene shallow-water sandstone with bryozoans and echinoids
	527	Middle Maastrichtian, deep-water sediments
	528	Maastrichtian, deep-water sediments
	529	Late Maastrichtian, deep-water sediments
	530	Albian or younger igneous rocks overlain by late Albian–Cenomanian red claystone
	531	Late Pleistocene sediments
	532	Late Miocene deep-water sediment
Falkland Plateau	327	Barremian, shallow-water fossils and reworked shells
	328	Palaeocene, deep-water sediments
	329	Palaeocene, deep-water sediments
	330	Jurassic coal

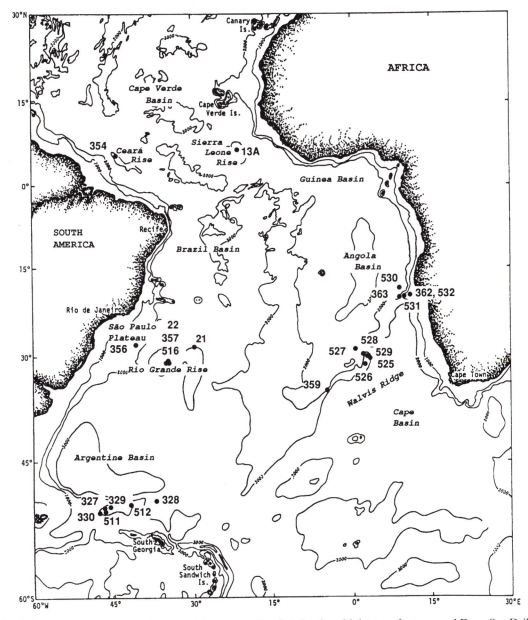

Fig. 2.1 Simplified bathymetric map of the present South Atlantic, with bottom features and Deep Sea Drilling Project sites discussed in the text. Contours are in fathoms.

Gulf of Guinea in the northern South Atlantic, the Rio Grande Rise and Walvis Ridge in the central South Atlantic, and the Falkland Plateau in the south. These relatively shallow areas were likely to have remained emergent longest as the South Atlantic widened and deepened. The northern sites, however, have not provided clues to the earliest history of the northern South Atlantic.

Southern South Atlantic

In the southern South Atlantic, six sites, four of them close together, were drilled on the Falkland Plateau (Fig. 2.1; Table 2.1). Sites 327, 330, and 511 penetrated older rocks. Site 327 reached Barremian-age rocks whose deposition in shallow water is indicated by the fossils and by the presence of reworked shell material. The oldest sediments were rich in organic material, possibly indicating deposition in a basin with restricted circulation (Barker *et al.* 1976). Marine organic-rich mudstones of Late Jurassic through early Albian Age were drilled at Site 511 (Ludwig *et al.* 1983). Drilling at Site 330 stopped in Palaeozoic or Precambrian metamorphic rocks overlain by Jurassic sediments that contain coal and are, therefore, of terrestrial origin. The Upper Jurassic and Lower Cretaceous rocks cored at Site 330 exhibit a well-developed sequence of non-marine to shallow marine sediments; this site probably lay below 200 m by the Cenomanian (Barker *et al.* 1976). Based on the timing of the opening of the South Atlantic determined from palaeomagnetic evidence by Larson and Ladd (1973) and the re-establishment of oxygenated water conditions at Site 327 and other sites in the South Atlantic, Barker *et al.* (1976) concluded that the eastern tip of the Falkland Plateau would have cleared the southern end of Africa at about the Aptian–Albian boundary.

Central South Atlantic: Walvis Ridge and Rio Grande Rise

The data from the northern and southern parts of the South Atlantic are consistent with Sclater *et al.*'s (1977) conclusion that the latest direct connection between South America and Africa was likely to have been in the central South Atlantic, along the Rio Grande Rise and the Walvis Ridge. The Rio Grande Rise and Walvis Ridge are now separated by the Mid-Atlantic Ridge and the deep areas on either side of it, but were once part of the same feature. The prevailing hypothesis is that the Rio Grande Rise and Walvis Ridge were formed by an unusual amount of volcanic material extruded from the rift (for example Thiede 1977; see Shaffer, 1984, for an alternative interpretation). In addition, they may be composed partly of continental crust (Dean *et al.* 1984). As the plates drifted apart and volcanic material continued to be extruded, the volcanic piles elongated into ridges. The same mechanism has produced Iceland, a volcanic pile that also sits astride the Mid-Atlantic Ridge. On the basis of Leg 4 deep-sea drilling project (DSDP) data, Benson *et al.* (1970) concluded that the Rio Grande Rise finally separated from the spreading ridge at about 100 mybp. Sclater *et al.*'s (1977) reconstruction for 95 mybp suggests that Benson *et al.*'s (1970) estimate might be early, and that the separation came between 95 and 80 mybp. Shaffer's (1984) interpretation of Walvis Ridge as an abandoned spreading axis does not counter this conclusion.

Five sites have been drilled on the São Paulo Plateau and Rio Grande Rise and eleven on Walvis Ridge (Table 2.1). Drilling at Site 21 reached a winnowed layer of shell debris that included the remains of red algae, which require light to grow. This layer was dated as Campanian. At Site 357, drilling reached middle Eocene sediments containing shallow water fossils and a volcanic breccia that probably was weathered subaerially. Drilling at the site closest to the Mid-Atlantic Ridge, Site 359 on Walvis Ridge, reached upper Eocene ash flow tuff that probably was extruded above sea-level (Supko *et al.* 1977). The five sites of the Leg 74 transect on Walvis Ridge (Moore *et al.* 1983) were closely spaced and spanned a present depth range from about 1100 m to about 4400 m. The shallowest, Site 526, bottomed in late Palaeocene sandstone containing fragments of oysters, echinoids, and shell fragments with borings of clionid sponges, which are all shallow-water indicators (Moore *et al.* 1983; Fütterer 1983). Site 363 penetrated as far as upper Aptian calcareous sand and algal material that was presumably derived from a nearby island (Bolli *et al.* 1978).

Thiede (1977) summarized the shallow-water sediments known from the Rio Grande Rise in an effort to better understand its emergence history. The samples that he considered included the Campanian shell debris from Site 21, the middle Eocene volcanic breccia from Site 357, and three dredge samples containing shallow-water limestones of Eocene, Oligocene, and Eocene and Cretaceous age. All five samples were taken from well below the crest of the Rise and were estimated by Thiede to have been deposited in water shallower than 100 m. The youngest samples are the shallowest on the Rise at present and the oldest are deepest. Because the samples all were deposited at approximately the same palaeodepths, with corrections for downslope movement, their present bathymetric spacing indicates the amount of subsidence of the Rise through time. The youngest sample, an Oligocene shallow-water limestone, now lies approximately 1600 m below sea-level, indicating that the Rise was about 1500 m higher as late as the Oligocene and the crest of the Rise was exposed. Thiede (1977) estimated that the Rio Grande Rise finally subsided below the surface of the ocean in the late Oligocene. Later drilling did not contradict Thiede's (1977) conclusions. Site 516 bottomed in Cretaceous sediments that were deposited in water as shallow as 20 m, and middle Eocene reefal debris containing calcareous algae and byrozoans occurred higher in the section (Barker *et al.* 1983).

The time of submergence of the Walvis Ridge can be estimated from DSDP data. Middle Eocene volcanics that were probably extruded above sea-level were drilled at Site 359 (Supko *et al.* 1977), well away from Africa on the western end of Walvis Ridge. The rocks now lie 1765 m below sea-level and about 200 m below the seafloor. The highest part of the ridge crest nearby

stands about 1100 m below sea-level. Site 359 is located on ocean floor that is 53–47 my old, as determined from the magnetic anomalies. This corresponds to the Eocene age of the drilled volcanics. The age of the marine sediments directly overlying the volcanics is late Eocene, so the part of the ridge containing Site 359 had subsided by then. Ocean crust is seen to subside very rapidly at first, as much as 500 m in 5–10 my. Therefore, the high part of the ridge near Site 359 was probably completely submerged by the end of the Eocene. The Eocene age of the volcanics implies that a significant height could not have been eroded from the ridge crest before submergence. Eocene faunas overlying the volcanics are considered to indicate 300–900 m palaeodepth (Schlanger 1981).

The most exhaustive compilations of subsidence curves of South Atlantic DSDP sites are those by Melguen (1978), Thierstein (1979), and Dean *et al.* (1984). Curves for the Rio Grande Rise are consistent with the conclusions reached by Thiede (1977), although Thierstein's (1979) curve for Site 364 begins at about 500 m depth 110 mybp, whereas Melguen (1978) indicated that the site was emergent at that time. Melguen's (1978) curves for the Falkland Plateau sites, which Thierstein (1979) did not reconstruct, present some problems in that she indicated that Site 329 was emergent 65–60 mybp, the approximate age of the relatively deep-water sediments at the bottom of the core. Further, according to her curves, at the time Site 329 was supposedly emergent, Site 330 lay about 1800 m below sea-level. Because Site 330 is now only 1100 m below Site 329, one must hypothesize extensive erosion or non-deposition at Site 329 and not at Site 330 to account for discrepancy. Such erosion has probably occurred locally in this area, although the erosion would have been accomplished by underwater currents (Ludwig and Krasheninnikov 1980). Despite these problems, the Falkland Plateau connection between South America and Africa evidently was severed earlier than either the Walvis Ridge–Rio Grande Rise or the Brazil–Gulf of Guinea connections farther north.

Unfortunately, subsidence curves on Walvis Ridge have not been constructed for Sites 359 and 526, which on palaeontological and sedimentological evidence appear to have been the shallowest (see previous discussion). Among the other sites, Site 525 was the shallowest at the beginning of its history. Dean *et al.* (1984, see their Fig. 13) estimated that the initial basalts were extruded at about 200 m water depth during a time of relative high sea-level stand in the Late Cretaceous.

Geology of the continents and continental margins

The margins of South America and Africa share many geological features (for example Torquato and Cordani 1981). Perhaps the two most striking similarities are the presence on both continents of Early Cretaceous sedimentary sequences and rift volcanics that are identical in lithology and age. North of the line described by the Rio Grande Rise and Walvis Ridge, continental margin basins containing the following sequences of Early Cretaceous rocks have been identified: river-deposited continental sediments overlain by thick deposits, which are in turn overlain by normal marine sediments deposited in increasingly deep water. In Brazil, the Serra Geral Basalts form a prominent suite of volcanics with a mean age of 127 my (Hauterivian age; Hertz 1966; Bigarella 1973; Campos *et al.* 1974). On the opposite side of the ocean, the Kaoko Lavas of South West Africa have a mean age of 125 my (Siesser 1978).

When the margins of South America and Africa are fitted in their probable pre-South Atlantic configuration (Bullard *et al.* 1965), Late Precambrian to Ordovician sediments in the Jaibara Basin of Brazil are aligned with those in the African Oti Basin, and coeval sediments in the Estancia Basin in Brazil and the Noya Basin in Africa are also aligned (Nairn and Stehli 1973; Campos *et al.* 1974). The Brazilian Precambrian mountain-building episode is represented by rocks on both sides of the South Atlantic that would have formed a continuous belt prior to the rift. Old continental cores, the Guiana and Western African Shields, are also contiguous when the South Atlantic is closed. Finally, the late Palaeozoic Parnaiba Basin appears to have had an extension in the Alto Volta Basin in Africa (Campos *et al.* 1974). In addition to the evidence from the rocks, terrestrial fossils provide abundant clues that Africa and South America were once joined as one continent (Cox 1974).

The earliest sea-floor magnetic anomalies, which are in the southern part of the South Atlantic (Ladd *et al.* 1973; Sclater *et al.* 1977; Austin and Uchupi 1982), and the rift volcanics on the continents are Neocomian in age and most of the pre-Aptian sedimentation on the edges of the continental masses apparently was terrestrial. The earliest widespread normal marine deposits from the northern South Atlantic region are Albian age (Asmus and Ponte 1973; Delteil *et al.* 1974; Evans 1978; Falkenhein *et al.* 1981), well into the opening of the South Atlantic, according to Sclater *et al.*'s (1977) reconstructions. This contradiction, with the earliest normal marine deposits on the continents appearing well after rifting, is only apparent. The continental margins would have been part of a topographic high along the initial rift, precluding marine deposition. However, the margins would have subsided by Albian time. Early Aptian deposits are evaporitic in the northern South Atlantic (Evans 1978). South of the Rio Grande Rise–Walvis Ridge barrier, however, normal marine conditions prevailed, as indicated by Aptian–Albian reef deposits in the Cuanza Basin (Franks and Nairn 1973). By the latest Aptian, even the northern South Atlantic basins, for example, the Sergipe–Alagoas Basin in easternmost Brazil, experienced some normal marine sedimentation (Asmus and Ponte 1973).

South American–African connections

Southern South Atlantic

Continental geology in southern South America and in South Africa gives few clues to the persistence of the land connection through the Falkland Plateau. Sclater *et al.*'s (1977) reconstructions indicate that, because of subsidence of the middle part of the plateau, the connection would have been severed even as the first rift basalts farther north were extruded, that is, earlier than Hauterivian time, although the tip of the submerged plateau did not clear South Africa until the Albian (Barker *et al.* 1976). Although circulation was restricted in the southern South Atlantic and evaporative conditions on the adjacent continents are indicated by the clays in the sediments (Natland 1978), no evidence exists that the basin was cut off from marine influence at any time from the Aptian onward.

Central South Atlantic

By Aptian–Albian time, the South Atlantic would have been well open, particularly in the southern part. Marine rocks of Early Cretaceous age are known from the Magallanes Basin in southern South America, from South Africa, and from Madagascar and East Africa (Natland *et al.* 1974). These rocks may represent marine incursions into the rift and further suggest that no barriers to marine incursions from the west or north (along East Africa) existed.

In addition to the widening of the basin by continental drift, the separation of South America and Africa by water was augmented by sea-level changes. According to Haq *et al.* (1987), global sea-level was at a low in the early Valanginian and thereafter rose steadily until the early Aptian. After a mid-Aptian regression, sea-level rose quickly through the Albian to a high level in the early Cenomanian. The result of this sea-level rise was the rapid widening of the seaway between South America and Africa. DSDP Site 363 (Walvis Ridge), which was in the centre of the seaway early in its history, contains marine sediments to the bottom of the core in the upper Aptian. Those sediments include algal material postulated to have come from a nearby island (Bolli *et al.* 1978), so the proximity of an emergent area is not ruled out. None the less, subsidence of the Walvis Ridge–Rio Grande Rise connection due

to crustal contraction coincided with a sea-level rise, and at least partial marine connections between the southern and northern South Atlantic must have been established by the late Aptian because the northern South Atlantic was marine, albeit restricted. As early as the Aptian, then, a full land connection between South America and Africa no longer existed across the central South Atlantic. Furthermore, with the exception of a drop in global sea-level in the mid-Cenomanian, which did not go to pre-Aptian levels, sea-level remained high throughout the Cretaceous and Palaeogene (Haq *et al.* 1987), with only one significant drop (still above present sea-level) at the Cretaceous–Tertiary boundary. The precise time when the Walvis Ridge–Rio Grande Rise axis was severed as a continuous land connection remains unknown, but probably was mid-Early Cretaceous. Island-hopping across widening water barriers would have been the only possible dispersal mode for organisms across the central South Atlantic after that time (Cox 1980).

Northern South Atlantic

A possible connection between South America and Africa across the northern South Atlantic is particularly difficult to evaluate. Relevant DSDP data do not exist, so that speculation about the presence of islands, such as persisted along the Walvis Ridge–Rio Grande Rise, is difficult. Continental geology is also unhelpful because some important regions are not well known. The three Brazilian basins of greatest concern are the Barreirinhas Basin, southeast of the Amazon River Delta, the Potiguar Basin on the northern side of the Brazilian 'nose', and the Recife–Joao Basin in easternmost Brazil. The following discussion is taken mostly from reviews by Asmus and Ponte (1973) and Ojeda (1982). The Barreirinhas Basin contains volcanics that are the same age (Hauterivian) as the rift volcanics farther south. Aptian marine rocks are present in this basin, but whether they are evaporitic, as are similar-age rocks in the southern basins, is not known (Asmus and Ponte 1973). A large-scale

Albian transgression, part of a global sea-level rise illustrated by Haq *et al.* (1987), flooded the basin.

Upper Albian rocks in the Barreirinhas Basin contain reefs and shelf limestones, indicating normal marine conditions. In contrast, Aptian–Albian rocks in the Potiguar Basin are non-marine, with the first transgression occurring at the beginning of the Cenomanian. The Cenomanian rocks are mostly continental, containing some littoral clastics, and the first fully marine rock is fossiliferous limestone of Turonian age. In the Recife–Joao Basin, the oldest known rocks are Turonian volcanics, overlain by littoral to terrestrial sediments containing faunas with affinities to those in Cameroon. Because the offshore geology of the Brazilian continental margin basins is poorly known, the presence of Cenomanian continental rocks does not necessarily indicate the existence of a land connection between South America and Africa at that time.

The African side of the equatorial Atlantic has a similar history, but the continental margin geology is better known. Machens (1973) reported marine rocks from boreholes in the Ivory Coast that are Albian and Cenomanian in age. These rocks include sequences of black shales containing ammonites. Delteil *et al.* (1974) also reported middle to late Albian marine sediments from the Ivory Coast–Ghana Ridge. These rocks place the initial opening of the Gulf of Guinea at Albian time.

The faunal evidence for a northern connection between South America and Africa is disputed, with some workers favouring a connection as young as Turonian across the Gulf of Guinea, in conflict with most of the geological data. Some of the faunal evidence was summarized by Reyment *et al.* (1976), who reviewed the Benue Trough and the Sergipe–Alagoas Basin faunas. The faunal data and the references cited by Reyment *et al.* (1976) include the following:

1. Terrestrial microfloras (Jardiné *et al.* 1974) formed one large province until the end of the Cenomanian, after which separate provinces on South America and Africa existed.

2. Niger Republic and South American dinosaurs were similar into the Coniacian (Buffetaut and Taquet 1975).
3. Pelecypods and gastropods were similar on both continents until the Cenomanian (Nicklés 1950). Freneix (1972) observed that Moroccan pelecypods of Turonian–Coniacian age were more similar to North American ones than to West African ones.
4. West African and Brazilian ostracodes showed strong affinities into the earliest Tertiary (Neufville 1973).

The importance of the data on terrestrial organisms is particularly subject to debate. Buffetaut and Taquet's (1975) conclusions about the similarities between South American and Africa dinosaur faunas was supported by the statistical analyses of Molnar (1980), but both studies showed that the faunas were not identical, and the similarities do not require a connection in the Late Cretaceous (Sues and Taquet 1979; also Buffetaut and Rage, Chapter 7 this volume). In addition, several studies have shown that hadrosaurs dispersed to South America from North America during the Late Cretaceous (Casamiquela 1980; Brett-Surman 1979). The absence of hadrosaurs in Africa is negative evidence for the lack of a connection with South America, but in light of their abundance and ubiquity, not entirely without value. Fresh-water fish taxa and the crocodilian, *Sarosuchus*, are common to South America and Africa in Aptian time but not later (Taquet 1978).

In addition to the faunal evidence cited by Reyment *et al.* (1976), Premoli-Silva and Boersma (1977) observed that a North Atlantic–South Atlantic connection was not indicated prior to Cenomanian time for foraminifera, although they pointed out that establishment of an apparent connection at that time could be due to a slight climatic change, rather than to a palaeogeographic change. Finally, Kennedy and Cooper (1975), Förster (1978), and Wiedmann and Neugebauer (1978) argued for a middle to late Albian connection between the North and South

Atlantic on the basis of ammonites, the first authors pointing out that much of the evidence for a post-Albian barrier is negative.

In evaluating the faunal evidence, one must take into account the oceanic currents that would have been present in the two halves of the ocean. Without presenting any justification for their models, Reyment *et al.* (1976) showed currents entering the gap between the North Atlantic and the South Atlantic from north and south. However, their reconstruction cannot be supported by dynamical arguments. The atmospheric circulation maps of Parrish and Curtis (1982), which can be used as a rough guide to the probable distribution of oceanic surface currents, suggest that the oceanic currents north and south of the equator would not have flowed through the gap but, rather, would have bypassed it and remained independent from each other, as is observed in the present-day ocean (see also reconstructions by Berggren and Hollister 1974). As soon as the ocean was wide enough to permit the development of a distinct equatorial current, the predominant flow would have been south to north because of the position of the Brazilian 'nose', which would have shunted water across the equator to the north just as it does now. In addition, Arthur and Natland (1979) argued for a haline current flowing south to north through the gap. Palaeontologists disagree on the evidence for south-to-north versus north-to-south migration of marine faunas. Förster (1978) claimed a predominance of north to south migrations among Albian ammonites and Reyment (1973) concluded that the reverse had occurred. The possibility that the high salinity of the South Atlantic could have been a barrier to marine migration is one that, to my knowledge, has not been addressed and cannot be ruled out. In any case, the possibility exists that the central Atlantic Ocean could have been a barrier to terrestrial organisms long before the marine faunas showed that the land connection had been broken. The currents described above would have assured that a flow of terrestrial organisms between South America and Africa by rafting, for example, was

likely to have been predominantly east to west rather than the reverse.

Thus, the weight of the palaeogeographical and geophysical data is on the side of a late Albian severance of the last connection between South America and Africa (see Buffetaut and Rage, Chapter 7 this volume). The faunal evidence does not strongly favour a later separation. Because the faunal evidence is conflicting, the equatorial Atlantic connection can be assumed to finally have severed at the close of the Early Cretaceous.

THE NORTH AMERICAN–SOUTH AMERICAN CONNECTION

Among the major plate tectonic events, the separation of South America and Africa is perhaps the simplest to understand. The history of the connections between North and South America is arguably the most complicated, with virtually every aspect the subject of controversy. The problem was neatly summed up by Case *et al.* (1984): '. . . all parts of the Caribbean region are allochthonous to variable degrees with respect to their neighbors . . .' A major stumbling block has been the fit of the continents in the Gulf of Mexico–Caribbean region. Without question, North and South America were in proximity because indisputable coeval connections existed between South America and Africa on the one hand and Africa and North America on the other. However, continental fits that do not violate the well-known relations among South America, Africa, and North America inevitably result in an overlap of South America by Mexico, Central America, and the Greater Antilles (Burke *et al.* 1984). Following is a list of the less debatable points:

1. Mexico, the Greater Antilles, and the Chortis Block (comprising Honduras and Nicaragua; Pindell and Dewey 1982; Burke *et al.* 1984), all contain rocks that predate the major separation of North and South America as recognized from the palaeomagnetic data. Few workers question that some or all of these fragments have

moved (Pindell and Barrett, 1990). The easiest movement to visualize is that of the Greater Antilles because major strike–slip faults and a well-defined spreading centre, the Cayman Trough, bound the Caribbean region. The questions centre on the timing of the movement.

2. Apart from basement metamorphics of the Chortis block, and with the possible exception of rocks on the Isle of Pines (Donnelly 1974), Palaeozoic rocks older than Late Permian have not been found in the Caribbean region. Rocks in the region are Jurassic and younger, with rare exceptions (intrusives expected from the collision of Gondwana and Laurussia, Late Permian of Guatemala, Dengo 1975; Triassic–Jurassic red beds, Yucatan, Lopez Ramos 1975 [believed to be Cretaceous by some workers, for example Donnelly 1975]). Banks (1975) also noted that the oldest rocks in the Greater Antilles are mostly volcanic. The age distribution of the rocks in the Gulf of Mexico and Caribbean is in contrast to the striking ubiquity of Palaeozoic and Precambrian rocks in the continental areas surrounding the region (Banks 1975). Thus, establishment of the history of connections between North and South America involves not only the timing of separation of the two plates, but also of the formation and history of Caribbean terranes.

3. Orogenic foldbelts—comprising the Appalachians, the Ouachita and Arbuckle Mountains, the Marathon and Llano uplifts in Texas, and the discontinuous Huastecan foldbelt in Mexico—border the entire eastern and southern portions of North America. These belts have been interpreted as a single, continuous belt (for example King 1975). The collision of North America and western Europe during Devonian and Early Carboniferous resulted in deformation in the northern and central Appalachians. However, all the other foldbelts experienced deformation in the Carboniferous and Permian, interpreted to have resulted from the collision of Laurussia and Gondwana.

Until recently, the major controversy that most confounded agreement on reconstructions of the

North American–South American connection concerned the fit of the continents around the Gulf of Mexico. Continental fits that are restored along the Atlantic sea-floor spreading magnetic anomalies are sufficiently uncertain that the varying degrees of overlap of Mexico and Central America with South America all are equally consistent with the Atlantic data (Burke *et al.* 1984). On the basis of continental palaeomagnetic data, for example, Van der Voo *et al.* (1976) proposed a 'tight fit' of North and South America, which fit northern South America into the Gulf of Mexico, and rotated all the intervening territory west of North America (also Dickenson and Coney 1980). The classic Pangaean reconstruction of Bullard *et al.* (1965), on the other hand, left a 'hole' at the Gulf of Mexico (they did not attempt to deal with the problem of Mexico, Central America, and the Greater Antilles). Van der Voo *et al.*'s (1976) reconstruction explained, among other features, the deformation in southern North America. Several workers (for example Cebull and Shurbet 1980) objected to the 'tight fit' because they felt that the Carboniferous deformation in southern North America, which created the Ouachita–Marathon foldbelt, was not sufficient to be the result of a major continental-continent collision, but those who objected to the 'tight fit' were left with a hole in the Gulf of Mexico across which the stress creating the deformation could not have been transmitted. In contrast to the deformation in North America, the collisional deformation in northern South America is severe, including tight folds and thrusts, batholiths, and decollement tectonics, and is also somewhat younger (Late Permian–Triassic; Shagam 1975; Irving 1975).

4. The argument that some sort of hole existed in the vicinity of the Gulf of Mexico has been encouraged by the presence of thick salt deposits that apparently are largely Jurassic in age. The salt occurs in two bodies, one along the United States Gulf Coast and one bordering the northern part of Yucatan (Buffler *et al.* 1980). The discontinuity of the salt deposit has been interpreted as

the consequence of post-salt-deposition rifting (Buffler *et al.* 1980; Pindell 1985), an interpretation that has been challenged because of a lack of well-defined magnetic anomalies (Cebull and Shurbet 1980).

Thus the separation of North and South America apparently involved movements that included rifting *sensu stricto*; alternating periods of compression and extension related to changes in plate motion as the Atlantic opened; extensive strike–slip faulting; subduction; and rotation of continental blocks. Unavoidably, a review of the palaeogeography of the Gulf of Mexico–Caribbean region will favour one interpretation over others, if only to provide some structure for presenting the data, and this review is no exception. The most recent, comprehensive, and coherent models of Caribbean plate evolution are those by Pindell and his co-workers (Pindell and Dewey 1982; Pindell *et al.* 1988; Pindell and Barrett 1990), which will be used as the foundation for the following discussion. Similar models have been presented by Dickinson and Coney (1980), Gose *et al.* (1980), and Anderson and Schmidt (1983), but these differ in some crucial details. Readers who are interested in the details of the models should refer to those papers. A comparison of 13 different models has been presented by Pindell and Barrett 1990.

Reconstruction of Pangaea in the Permian, before the initiation of rifting, has the following features:

1. Yucatan is rotated clockwise relative to its present position and fitted against the United States Gulf Coast. This reconstruction solves two problems. It explains the distribution of the salt, which partly occupies the Yucatan block and was divided postdepositionally, apparently by seafloor spreading (Buffler *et al.* 1980). The reconstruction also solves the problem of relatively weak deformation in the foldbelts along the Gulf Coast (for example Walper 1980; Pindell and Dewey 1982) by placing an intervening block in front of them that would have absorbed much of the impact of Gondwana. The Palaeozoic basement rocks of Yucatan are highly deformed (see Pindell 1985, and references therein) and apparently were metamorphosed in the Early Carboniferous (Lopez Ramos 1975), consistent with the timing of the collision between Gondwana and North America. Pindell (1985) included the Florida Straits block in the 'hole', to occupy the space west of the Florida peninsula, and further tightened the fit by restoring crust attenuated by rifting and extension to its original thickness. Alternate interpretations rotate the Yucatan block (called the Maya East Block in some papers) clockwise in place, with no other movement relative to the North American craton, or show no rotation (for example Dickinson and Coney 1980; Gose *et al.* 1980; Anderson and Schmidt 1983).

2. Mexico is divided into three east–west segments along shear zones and rotated westward adjacent to the present west coast of the United States (Anderson and Schmidt 1983). The Chortis Block also is rotated westward, along the present-day Motagua Fault zone. The eastern side of Mexico lay, then, along the north-western coast of South America. The Mexican segment nearest the craton is rotated the least, relative to its present position with respect to the craton, and the Chortis Block is rotated farthest. The Motagua Fault is presently active, but the zone also contains older ophiolites, indicating a prolonged history of tectonic activity (Burke *et al.* 1984; Pindell and Barrett 1990). The northern shear zone in the reconstruction of Mexico is the Monterrey–Torreon Shear zone (Silver and Anderson 1974; de Cserna 1976). The primary evidence for the two remaining shear zones is apparent offset in the Huastecan foldbelt, but the poor exposure of the belt precludes a firmer interpretation. Thus, Gose *et al.* (1980) and Dickinson and Coney (1980) recognized only one shear, approximately at the position of the Monterrey–Torreon Shear Zone.

The opening of the Gulf of Mexico

Tectonic movements between North and South America began with the initiation of rifting in the North Atlantic. The North Atlantic opened along the former suture between North America and Africa, and Pindell and Dewey (1982) proposed a continuation of the North Atlantic rift system parallel to the former suture between North and South America and north of the Yucatan Block. As the North Atlantic widened, the Gulf of Mexico opened, creating a restricted basin that filled with salt (Walper 1980). Continued opening of the Gulf of Mexico along a spreading centre, with Yucatan rotating counterclockwise as it moved south, split the salt into two bodies, the Louann Salt of the United States Gulf Coast and the salts of eastern Mexico and Campeche Bank (Buffler *et al.* 1980; Pindell and Dewey 1982; Pindell 1985).

Left-lateral movement along the shear zones in Mexico and north of the Chortis Block moved the Chortis Block and the pieces of Mexico into their present positions relative to the North American craton. Pindell and Dewey (1982) suggested that the opening of the Gulf of Mexico and the emplacement of Mexico and Yucatan (but not the Chortis Block, which is still moving today) were essentially complete in the Early Cretaceous. During this phase of the relative movement between North and South America, a continuous land bridge could have existed through Yucatan, which according to Pindell and Dewey's (1982) model, was in contact with both continents. Both Dickinson and Coney's (1980) and Anderson and Schmidt's (1983) reconstructions place Yucatan farther north.

The early history of the Caribbean

As spreading ceased in the Gulf of Mexico, the continued separation of North America and Gondwana was probably accomplished by spreading between Yucatan and northern South America (Fox and Heezen 1975; Mooney 1980; Dickinson and Coney 1980; Pindell and Dewey 1982). This spreading could have severed any land connection between North and South America as early as the Tithonian (latest Jurassic; Pindell and Barrett 1990; Ross and Scotese 1988). Pindell and Dewey (1982) hypothesized that in the Pacific the spreading ridge terminated in a fracture zone. The fracture zone would have been a site of crustal weakness, which evolved into a subduction zone dipping southwestward. The magmatism associated with the subduction created an island arc, which eventually was to become the core of the Greater Antilles (Dickinson and Coney 1980) and Aves Ridge (Pindell *et al.* 1988, Fig. 4 therein). By early Late Cretaceous time, this arc, which collided with both Yucatan and northern South America, may have bridged the gap between the two continents (Pindell and Barrett 1990, Plate 1c therein). The island arc proceeded to migrate northeastward over a north-east-facing subduction zone, which consumed the spreading ridge to the east (this was extinct by early Campanian time; Pindell *et al.* 1988) and the oceanic crust that the ridge had created; the southwest-facing subduction zone ceased. Thus pre-existing Pacific Ocean crust was intercalated between North America and South America behind the migrating arc.

In its north-eastern movement, the western end of the Antillean arc collided with southern Yucatan. The result of this collision was to begin the breakup of the arc into separate blocks along shear zones resulting from the collision and along the shear zones bounding either side of the migrating Caribbean plate. These separate blocks eventually constituted Jamaica, the southern Haitian peninsula, and two central blocks of Hispaniola, one of which also contained Puerto Rico.

Subduction was initiated along the eastern end of the Antillean arc, probably partly induced by the greater eastward component of the movement of the arc that resulted from the collision with Yucatan. This subduction raised a chain of now-extinct volcanoes in the present Aves Ridge, which lies just west of the Lesser Antilles (Fox and Heezen 1975). Also during this time, a major magmatic event, whose significance remains enigmatic, created the basaltic B″ horizon (named for its designation in seismic reflection profiles; Case 1975). The B″ horizon, which is probably a sill, blankets the Caribbean plate and has been dated as Coniacian; the deepest DSDP holes in the Caribbean have terminated in this basalt (Edgar *et al.* 1973).

During the Late Cretaceous, a new subduction zone was initiated in the Pacific, west of the evolving Caribbean. The magmatic arc associated with this subduction eventually would become the Isthmus of Panama. Throughout the Late Cretaceous, the Caribbean plate continued to move north-eastward, further fragmenting the arc that was evolving into the Greater Antilles. Near the close of the Cretaceous, back-arc spreading split the Antillean arc lengthwise, opening the Yucatan Basin. The southern part of the split block constituted what is now the Cayman Ridge and the southern peninsula of Haiti. North of the rift lay what was to become western Cuba.

The Eocene was a time of major tectonic and volcanic activity in the Caribbean Sea. In the early Eocene, the northern part of the Antillean arc collided with the Blake–Bahama Platform, shutting off the south-dipping subduction zone along which the arc had been migrating. From that time on, the motion of the Caribbean plate was primarily eastward, with west-dipping subduction and accompanying volcanism along the Lesser Antilles. The volcanism in the Lesser Antilles was extremely vigorous in the Eocene and has been decreasing ever since (Tomblin 1975).

The land connection between North America and South America

One of the difficulties in understanding the early history of the connection between North and South America is a gap in the record for much of the Early Cretaceous (Pindell and Barrett 1990; Ross and Scotese 1988). The oldest rocks penetrated in the Caribbean by DSDP drilling are the Coniacian-age B″ horizon, which are not the true oceanic basement (Case 1975), so that evidence of the pre-B″-horizon history cannot be inferred yet from the evolution of sediment patterns as the plate passed between North and South America, and critical information about the connections among the evolving islands is lacking. Nevertheless, some constraints are provided by the plate motions and by what little geology is known.

According to Pindell and Dewey (1982) and Pindell (1985), the motions of the opening Gulf of Mexico during the Jurassic were partly taken up by strike–slip faults between western Yucatan and Mexico and southern Yucatan and northern South America. A marine connection between the Central Atlantic and the Pacific is indicated by relatively free faunal exchange in the Toarcian to early Bajocian (Hallam 1983). This does not necessarily imply separation of continental crust, however, but may have occurred during a highstand of sea level (Hallam 1983). Although Pacific Jurassic ammonites are now found in western Cuba, western Cuba's location in the Jurassic is not known and the ammonites are not diagnostic. It is reasonable to assume that North and South America were connected by land through the Chortis Block and Yucatan certainly through the Middle Jurassic and possibly until the beginning of the Cretaceous (Dickinson and Coney 1980; Anderson and Schmidt 1983; Pindell 1985).

By the start of the Cretaceous, Mexico was in place and North and South America were no longer connected through Yucatan, owing to the initiation of spreading. Intermittent connections through the developing Antillean arc were possible. During the Late Cretaceous, major orogenic activity in the Sierra Madre Oriental occurred and was associated with migration of the arc at the leading edge of the Caribbean plate (Pindell 1985). By the Campanian, and possibly as early as the Cenomanian (Pindell and Barrett 1990), the western end of the Antillean arc had collided with Yucatan, and the Aves Swell, which was connected to South America, had developed at the eastern end of the arc. Unfortunately, the results of drilling in the nearest DSDP sites, Sites 30 and

146, are not helpful in understanding possible connections. A connection between North and South America through Yucatan, the Antillean arc, and the Aves Swell during the Late Cretaceous is supported by the geology of the various islands, which includes continental and very shallow marine deposits. Indeed, a connection at this time is required by the Late Cretaceous migration of hadrosaurs into South America from North America.

During the Early Tertiary, the Yucatan Basin opened and in the Eocene, the Antillean arc collided with the Blake-Bahama Plateau. In addition, abundant subaerial volcanics of Eocene age were extruded on the Lesser Antilles (Tomblin 1975) and an Eocene? subaerial weathering surface (Edgar, et al. 1973) and shallow-water carbonates (Fox and Heezen 1975) are found on Aves Ridge. These geological features suggest that a connection between North America and South America most likely existed during Eocene time. The length of time this connection persisted is not known, but the Lesser Antillean volcanism decreased after the Eocene and many parts of the Caribbean subsided several hundred metres beginning in the Miocene. For example, marine rocks indicative of considerable Miocene and post-Miocene subsidence are found on Aves Ridge and north of Puerto Rico (Fox and Heezen 1975; Tomblin 1975). At the same time, however, the Panama arc, which was already in contact with North America, collided with northern South America (Pindell and Barrett 1990), so connections via that route were possible.

In summary, terrestrial connections between North and South America most likely existed through most of the Jurassic, with occasional interruptions at times of high sea-level, Late Cretaceous, and Eocene.

NORTH AMERICAN–AFRICAN CONNECTION

Severance of the connection between South America and North America at the Jurassic–Cretaceous boundary also did away with any connection between North America and Africa through South America. By the time of the Late Cretaceous connection between North and South America, the connection between South America and Africa was no more.

Connections between North America and Africa through Europe hinge on interpretations of the evolution of the Mediterranean, which is a remnant of the Tethyan seaway. The opening of the Central Atlantic during the Jurassic constituted an extension of the western end of the Tethyan seaway, which also was being rifted. By the late Jurassic, both the North Atlantic and Tethys were wide enough to prevent effective dispersal of terrestrial vertebrates. The opening of the Central Atlantic Ocean and Tethys created a southward motion of Gondwana relative to Laurasia, the northern continent consisting of North America and Eurasia. This sense of movement was reversed when the South Atlantic opened and Africa began moving northward toward Europe. The closing of Tethys between Africa and Europe has a complicated history, consisting of several cycles of collision and rifting. For example, Apulia (Italy, Yugoslavia, and parts of Hungary and Rumania) collided with Europe in the Late Cretaceous (Burchfiel and Royden 1982), but was simultaneously rifted from northern Africa. A brief connection between Europe and Africa was possible then. Additional land connections between Africa and Europe were probably not established until Oligocene or Miocene time. By the Miocene, the Mid-Atlantic Ridge had extended through to the Arctic Ocean (Pitman et al. 1974), effectively cutting off any connection between North America and Europe. Connections between North America and Africa were likely to have existed during the Late Cretaceous and during the Oligocene, when Africa was near Europe and when relatively shallow land connections still existed between Greenland and Europe across Spitsbergen and the Rockall Plateau. Connections were likely to have been sporadic or non-existent during most of the Tertiary (see further discussion in Parrish 1987).

SUMMARY

Connections between South America and Africa probably persisted longest along the Rio Grande Rise–Walvis Ridge, although constraints on the timing of separation in the Gulf of Guinea region are few. Most palaeogeographical and biogeographical data are consistent with a final separation sometime in the Albian. By contrast, connections between North and South America probably persisted, at least intermittently, into the Tertiary. Available data suggest connections through the Jurassic, in the Late Cretaceous, and in the Eocene. Connections between Africa and North America via South America ended with the separation of North and South America around the end of the Jurassic. By the time a connection between North and South America was re-established in the Late Cretaceous, South America and Africa had separated, and connection between North America and Africa was possible only via western Europe. This route was probably available for terrestrial dispersal in the Late Cretaceous and Oligocene.

REFERENCES

Anderson, T. H. and Schmidt, V. A. (1983). The evolution of Middle America and the Gulf of Mexico–Caribbean Sea region during Mesozoic time. *Geological Society of America Bulletin*, **94**, 941–66.

Arthur, M. A. and Natland, J. H. (1979). Carbonaceous sediments in the North and South Atlantic: the role of salinity in stable stratification of Early Cretaceous basins. In *Deep drilling results in the Atlantic Ocean: Continental margins and paleoenvironment* (ed. M. Talwani, W. Hay, and W. B. F. Ryan), Maurice Ewing Series 3, pp. 375-401.

Asmus, H. E. and Ponte, F. C. (1973). The Brazilian marginal basins. In *The ocean basins and margins*, Vol. 1, The South Atlantic (ed. A. E. M. Nairn and F. G. Stehli), pp. 87–133. Plenum Press, New York.

Austin, J. A., Jr. and Uchupi, E. (1982). Continental-oceanic crustal transition off Southwest Africa. *American Association of Petroleum Geologists Bulletin*, **66**, 1328–47.

Banks, P. O. (1975). Basement rocks bordering the Gulf of Mexico and the Caribbean Sea. In *The ocean basins and margins*, Vol. 3, The Gulf of Mexico and the Caribbean (ed. A. E. M. Nairn and F. G. Stehli), pp. 181–99. Plenum Press, New York.

Barker, P. F., Carlson, R. L., and Johnson, D. A. (1983). *Initial Reports of the Deep Sea Drilling Project*, Vol. 72. US Government Printing Office, Washington, DC.

Barker, P. F., Dalziel, I. W. D., *et al.* (1976*a*). Site Reports. In *Initial Reports of the Deep Sea Drilling Project*, Vol. 36 (ed. P. F. Barker *et al.*). pp. 17–266. US Government Printing Office, Washington, DC.

Barker, P. F., Dalziel, I. A. W., *et al.*, with Harris, W. and Sliter, W. V. (1976*b*). Evolution of the southwestern Atlantic Ocean Basin: Results of Leg 36, Deep Sea Drilling Project. In *Initial Reports of the Deep Sea Drilling Project*, Vol. 36 (ed. P. F. Barker *et al.*). US Government Printing Office, Washington, DC.

Benson, W. E., Gerard, R. D., and Hay, W. W. (1970). Summary and conclusions. In *Initial Reports of the Deep Sea Drilling Project*, Vol. 4 (R. G. Gerard, R. D. Gerard, *et al.*), pp. 659–73. US Government Printing Office, Washington, DC.

Berggren, W. A. and Hollister, C. D. (1974). Paleography, paleobiogeography and the history of circulation in the Atlantic Ocean. In *Studies in paleoceanography* (ed. W. W. Hay), Society of Economic Paleontologists and Mineralogists Special Publication **20**, pp. 126–86.

Bigarella, J. J. (1973). Geology of the Amazon and Parnaiba Basins. In *The ocean basins and margins*, Vol. 1, The South Atlantic (ed. A. E. M. Nairn and F. G. Stehli), pp. 25–86. Plenum Press, New York.

Bolli, H. M., Ryan, W. B. F., *et al.* (1978). *Initial Reports of the Deep Sea Drilling Project*, Vol. 40. US Government Printing Office, Washington, DC.

Brett-Surman, M. K. (1979). Phylogeny and paleobiogeography of hadrosaurian dinosaurs. *Nature*, **277**, 560–2.

Buffetaut, E. and Taquet, P. (1975). Les vertèbrés du Crétacé et la dérive des continents. *La Recherche*, **55**(6), 379–81.

Buffler, R. T., Watkins, J. S., Shaub, F. J., and Worzel, J. L. (1980). Structure and early geologic history of the deep central Gulf of Mexico Basin. In *The origin of the Gulf of Mexico and the early opening of the central North Atlantic Ocean* (ed. R. H. Pilger), pp. 3–16. Proceedings of a symposium at Baton Rouge, Louisiana State University.

Bullard, E. C., Everett, J. E., and Smith, A. G. (1965).

The fit of the continents around the Atlantic. *Philosophical Transactions of the Royal Society of London*, Series A, **258**, 41–51.

Burchfiel, B. C. and Royden, L. (1982). Carpathian Foreland fold and thrust belt and its relation to Pannonian and other basins. *American Association of Petroleum and Geologists Bulletin*, **66**, 1179–95.

Burke, K., Cooper, C., Dewey, J. F., Mann, P., and Pindell, J. L. (1984). Caribbean tectonics and relative plate motions. In *The Caribbean–South American plate boundary and regional tectonics* (ed. W. E. Bonini, R. B. Hargraves, and R. Shagam), *Geological Society of America Memoir*, **162**, 31–63.

Campos, C. W. M., Ponte, F. C., and Miura, K. (1974). Geology of the Brazilian continental margin. In *The geology of continental margins* (ed. C. A. Burk and C. L. Drake), pp. 447–61. Springer–Verlag, New York.

Casamiquela, R. M. (1980). Considérations écologiques et zoogéographiques sur les vertèbrés de la zone littorale de la mer du Maestrichtien dans le Nord de la Patagonie. *Memoires Société Géologique France*, NS **139**, 53–5.

Case, J. E. (1975). Geophysical studies in the Caribbean Sea. In *The ocean basins and margins*, Vol. 3, The Gulf of Mexico and the Caribbean (ed. A. E. M. Nairn and F. Stehli), pp. 107–80. Plenum Press, New York.

Case, J. E., Holcombe, T. L., and Martin, R. G. (1984). Map of geologic provinces in the Caribbean region. In *The Caribbean–South American plate boundary and regional tectonics* (ed. W. E. Bonini, R. B. Hargraves, and R. Shagam), *Geological Society of America Memoir*, **162**, 1–30.

Cebull, S. E. and Shurbet, D. H. (1980). The Ouachita Belt in the evolution of the Gulf of Mexico. In *The origin of the Gulf of Mexico and the early opening of the central North Atlantic Ocean* (ed. R. H. Pilger), pp. 17–26. Proceedings of a symposium at Baton Rouge, Louisiana State University.

Cox, C. B. (1974). Vertebrate paleodistributional patterns and continental drift. *Journal of Biogeography*, **1**, 75–94.

Cox, C. B. (1980). An outline of the biogeography of the Mesozoic world. *Mémoires Société Géologique France*, NS **139**, 75–9.

de Cserna, Z. (1976). Mexico–geotectonics and mineral deposits. *New Mexico Geological Society Special Publication*, **6**, 18–25.

Dean, W. E., Hay, W. W., and Sibuet, J.-C. (1984).

Geologic evolution, sedimentation, and paleo-environments of the Angola Basin and adjacent Walvis Ridge: Synthesis of results of Deep Sea Drilling Project Leg 75. *Initial reports of the Deep Sea Drilling Project*, Vol. 75 (ed. W. W. Hay *et al.*), pp. 509–44. US Government Printing Office, Washington, DC.

Delteil, J. R., Valery, P., Montadert, L., Fondeau, C., Patriat, P., and Mascle, J. (1974). Continental margin in the northern part of the Gulf of Guinea. In *The geology of the continental margins* (ed. C. A. Burk and C. L. Drake), pp. 197–311. Springer–Verlag, New York.

Dengo, G. (1975). Paleozoic and Mesozoic tectonic belts in Mexico and Central America. In *The ocean basins and margins*, Vol. 3, The Gulf of Mexico and the Caribbean (ed. A. E. M. Nairn and F. G. Stehli), pp. 283–323. Plenum Press, New York.

Dickinson, W. R. and Coney, P. J. (1980). Plate tectonic constraints on the origin of the Gulf of Mexico. In *The origin of the Gulf of Mexico and the early opening of the central North Atlantic Ocean* (ed. R. H. Pilger), pp. 27–36. Proceedings of a symposium at Baton Rouge, Louisiana State University.

Donelley, T. W. (1975). The geological evolution of the Caribbean and the Gulf of Mexico. In *The ocean basins and margins*, Vol. 3, The Gulf of Mexico and the Caribbean (ed. A. E. M. Nairn and F. G. Stehli), pp. 663–89. Plenum Press, New York.

Edgar, N. T., Saunders, J. B., *et al.* (1973). Site 148. In *Initial reports of the Deep Sea Drilling Project*, Vol. 15 (ed. N. T. Edgar *et al.*), pp. 217–75. US Government Printing Office, Washington, DC.

Evans, R. (1978). Origin and significance of evaporites in basins around Atlantic margins. *American Association of Petroleum Geologists Bulletin*, **62**, 223–34.

Falkenhein, F. U. H., Franke, M. R., and Carozzi, A. V. (1981). Petroleum geology of the Macaé Formation (Albian–Cenomanian), Campos Basin, Brazil. *Ciencia, Técnica, Petróleo Secao: Exploracao de Petróleo*, **11**, 140 p.

Förster, R. (1978). Evidence for an open seaway between northern and southern proto-Atlantic in Albian times. *Nature*, **272**, 158–9.

Fox, P.J. and Heezen, B. C. (1975). Geology of the Caribbean crust. In *The ocean basins and margins*, Vol. 3, The Gulf of Mexico and the Caribbean (ed. A. E. M. Nairn and F. G. Stehli), pp. 421–66. Plenum Press, New York.

Franks, S. and Nairn, A. E. M. (1973). The equatorial marginal basins of West Africa. In *The ocean basins and margins*, Vol. 1, the South Atlantic (ed. A. E. M. Nairn and F. G. Stehli), pp. 301–50. Plenum Press, New York.

Freneix, S. (1972). Le bassin côtier de Tarfaya (Maroc Méridional). Tom III. Paléontologie. Les Mollusques bivalves crétacés du bassin côtier de Tarfaya (Maroc Méridional). *Notes Mémoires Service Géologique Maroc*, **228**, 49-255.

Fütterer, D. K. (1983). Evidence of clionid sponges in sediments of the Walvis Ridge, southeastern Atlantic site 526, Deep Sea Drilling Project, Leg 74. In *Initial reports of the Deep Sea Drilling Project*, Vol. 74 (ed. T. C. Moore *et al.*), pp. 557–9. US Government Printing Office, Washington, DC.

Gose, W. A., Scott, G. R., and Swartz, D. K. (1980). The aggregation of Mesoamerica: paleomagnetic evidence. In *The origin of the Gulf of Mexico and the early opening of the central North Atlantic Ocean* (ed. R. H. Pilger), pp. 51–4. Proceedings of a symposium at Baton Rouge, Louisiana State University.

Hallam, A. (1983). Early and mid-Jurassic molluscan biogeography and the establishment of the central Atlantic seaway. *Palaeogeography, Palaeoclimatology, Palaeoecology*, **43**, 181–93.

Haq, B., Hardenbol, J., and Vail, P. R. (1987). Chronology of fluctuating sea level since the Triassic (250 million year to present). *Science*, **235**, 1156–67.

Hartland, W. B., Cox, A. V., Llewellyn, P. G., Pickton, C. A. G., Smith, A. G., and Walters, R. (1982). *A geologic time scale*. Cambridge University Press, Cambridge.

Herz, N. (1966). Tholeiitic and alkalic volcanism in southern Brazil. *International Field Institute Guidebook*, Brazil, pp. V–1–V–6. American Geological Institute, Washington, DC.

Irving, E. M. (1975). Structural evolution of the northernmost Andes, Colombia. *US Geological Survey Professional Paper*, **846**,

Jardiné, S., Kieser, G., and Reyre, Y. (1974). L'individualisation progressive du continent africain vue à travers les données palynologiques de l'ère secondaire. *Bull. Sci. Géol. Strasbourg*, **27**, 69–85.

Kennedy, W. J. and Cooper, M. (1975). Cretaceous ammonite distributions and the opening of the South Atlantic. *Journal of the Geological Society of London*, **131**, 283–8.

King, P.B. (1975). The Ouachita and Appalachian orogenic belts. In *The ocean basins and margins*, Vol. 3, The Gulf of Mexico and the Caribbean (ed. A. E. M. Nairn and F. G. Stehli), pp. 201–41. Plenum Press, New York.

Klitgord, K. D. and Grow, J. A. (1980). Jurassic seismic stratigraphy and basement structure of the western North Atlantic quiet zone. *American Association of Petroleum Geologists Bulletin*, **64**, 1658–80.

Ladd, J. W., Dickson, G. O., and Pitman, Walter C., III (1973). The age of the South Atlantic. In *The ocean basins and margins*, Vol., The South Atlantic (ed. A. E. M. Nairn and F. G. Stehli), pp. 555–73. Plenum Press, New York.

Larson, R. L. and Ladd, J. (1973). Evidence for the opening of the South Atlantic in the Early Cretaceous. *Nature*, **246**, 209–12.

Lopez Ramos, E. (1975). Geological summary of the Yucatan Peninsula. In *The ocean basins and margins*, Vol. 3, The Gulf of Mexico and the Caribbean (ed. A. E. M. Nairn and F. G. Stehli), pp. 257–82. Plenum Press, New York.

Ludwig, W. J. and Krasheninnikov, V. (1980). Tertiary and Cretaceous paleoenvironments in the southwest Atlantic Ocean: preliminary results of Deep Sea Drilling Project Leg 71. *Geological Society of America Bulletin*, **Part I 91**, 655–64.

Ludwig, W. J., Krasheninnikov, V., *et al.* (1983). *Initial Reports of the Deep Sea Drilling Project*, Vol. 71. US Government Printing Office, Washington, DC.

Machens, E. (1973). The geologic history of the marginal basins along the north shore of the Gulf of Guinea. In *The ocean basins and margins*, Vol. 1, The South Atlantic (ed. A. E. M. Nairn and F. G. Stehli), pp. 351–90. Plenum Press, New York.

Melguen, M. (1978). Facies evolution, carbonate dissolution cycles in sediments from the eastern South Atlantic (DSDP Leg 40) since the Early Cretaceous. In *Initial reports of the Deep Sea Driling Project*, Vol. 40 (ed. H. M. Bolli *et al.*), pp. 981–1024. US Government Printing Office, Washington, DC.

Molnar, R. (1980). Australian late Mesozoic terrestrial tetrapods: some implications. *Mémoires Société Géologique France*, NS **139**, 131–43.

Mooney, W. D. (1980). An East Pacific–Caribbean Ridge during the Jurassic and Cretaceous and the evolution of western Colombia. In *The origin of the Gulf of Mexico and the early opening of the central North Atlantic Ocean* (ed. R. H. Pilger), pp. 55–74. Proceedings of a symposium at Baton Rouge, Louisiana State University.

Moore, T. C., Jr., Rabinowitz, P. D., Borella, P., Boersma, A., and Shackleton, N. J. (1983). Introduction and explanatory notes. In *Initial Reports of the Deep Sea Drilling Project*, Vol. 74 (ed. T. C. Moore *et al.*), pp. 3–39. US Government Printing Office, Washington, DC.

Nairn, A. E. M. and Stehli, F. G. (1973). A model for the South Atlantic. In *The ocean basins and margins*, Vol. 1, The South Atlantic (ed. A. E. M. Nairn and F. G. Stehli), pp. 1–24. Plenum Press, New York.

Natland, J. H. (1978). Composition, provenance, and diagenesis of Cretaceous clastic sediments drilled on the Atlantic continental rise off southern Africa, DSDP Site 361—implications for the early circulation of the South Atlantic. In *Initial Reports of the Deep Sea Drilling Project*, Vol. 40 (ed. H. M. Bolli *et al.*), pp. 1025–61. US Government Printing Office, Washington, DC.

Natland, M. L., Gonzales, E., Canon, A., and Ernst, M. (1974). A system of stages for correlation of Magallanes Basin sediments. *Geological Society of America Memoir*, **139**, 126 p.

Neufville, E. M. H. (1973). Upper Cretaceous–Palaeogene Ostracoda from the South Atlantic. *Publ. Paleont. Inst. Uppsala Spec.*, Vol. **1**, 205 p.

Nicklés, M. (1950). Mollusques testaces marins de la côte occidentale d'Afrique. *Manuels Ouest-Africains*, Vol. 2. Lechevalier, Paris.

Ojeda, H. A. O. (1982). Structural framework, stratigraphy, and evolution of Brazilian marginal basin. *American Association of Petroleum Geologists Bulletin*, **66**, 732–49.

Parrish, J. M., Parrish, J. T., and Ziegler, A. M. (1986). Permian-Triassic paleogeography and paleoclimatology and implications for therapsid distribution. In *The biology and ecology of mammal-like reptiles* (ed. J. Roth, C. Roth, and N. Hotton III). Smithsonian Press, Washington DC.

Parrish, J. T. (1987). Global palaeogeography and palaeoclimate of the Late Cretaceous and Early Tertiary. In *The origins of angiosperms and their biological consequences* (ed. E. M. Friis, W. G. Chaloner, and P. R. Crane), pp. 51–73. Cambridge University Press, Cambridge.

Parrish, J. T. and Curtis, R. L. (1982). Atmospheric circulation, upwelling, and organic-rich rocks in the Mesozoic and Cenozoic. *Palaeogeography, Palaeoclimatology, Palaeoecology*, **40**, 31-66.

Pindell, J. L. (1985). Alleghenian reconstruction and the subsequent evolution of the Gulf of Mexico, Bahamas and proto-Caribbean Sea. *Tectonics*, **4**, 1–39.

Pindell, J. L. and Barrett, S. F. (1990). Geological evolution of the Caribbean region: A plate-tectonic perspective. In *Decade of North American geology Caribbean region*, Vol. H (ed. J. E. Case and G. Dengo), pp. 405–32. Geological Society of America, Boulder, Colorado.

Pindell, J. L., Cande, S. C., Pitman, W. C., III, Rowley, D. B., Dewey, J. F., LaBrecque, J., and Haxby, W. (1988). A plate-kinematic framework for models of Caribbean evolution. *Tectonophysics*, **155**, 121–38.

Pindell, J. and Dewey, J. F. (1982). Permo–Triassic reconstruction of western Pangaea and the evolution of the Gulf of Mexico/Caribbean region. *Tectonics*, **1**, 1179–211.

Pitman, W. C., III, Larson, R. L., and Herron, E. M. (1974). The age of the ocean basins. *Geological Society of America Map and Chart Series*, **MC-6**.

Premoli-Silva, I. and Boersma, A. (1977). Cretaceous planktonic foraminifers—DSDP Leg 39 (South Atlantic). In *Initial Reports of the Deep Sea Drilling Project*, Vol. 39 (ed. P. R. Supko *et al.*), pp. 615–41. US Government Printing Office, Washington, DC.

Rabinowitz, P. D. and LaBrecque, J. (1979). The Mesozoic South Atlantic Ocean and evolution of its continental margins. *Journal of Geophysical Research*, **84**, 5973–60002.

Reyment, R. A. (1973). Cretaceous history of the South Atlantic Ocean. In *Implications of continental drift for the earth sciences* (ed. D. H. Tarling and S. K. Runcorn), pp. 805–14. Academic Press, London.

Reyment, R. A., Pengtson, P., and Tait, E. A. (1976). Cretaceous transgressions in Nigeria and Sergipe-Alagoas (Brazil). *An. Acad. Bras. Cience.*, **48** (Suppl.), 253–64.

Ross, M. I. and Scotese, C. R. (1988). A hierarchical tectonic model of the Gulf of Mexico and Caribbean region. *Tectonophysics*, **155**, 139–68.

Schlanger, S. I. (1981). Shallow-water limestones in oceans basins as tectonic and paleoceanic indicators. In *The deep-sea drilling project: a decade of progress* (ed. J. E. Warme, R. G. Douglas, and E. L. Winterer), *Society of Economic Paleontologists and Mineralogists Special Publication*, **32**, 209–26.

Sclater, J. G., Hellinger, S., and Tapscott, C. (1977). The paleobathymetry of the Atlantic Ocean from the Jurassic to the present. *Journal of Geology*, **85**, 509–52.

Shaffer, F. R. (1984). The origin of the Walvis Ridge:

sediment/basalt compensation during crustal separation. *Palaeogeography, Palaeoclimatology, Palaeoecology*, **45**, 87–100.

Shagam, R. (1975). The northern termination of the Andes. In *The ocean basins and margins*, Vol. 3, The Gulf of Mexico and the Caribbean (ed. A. E. M. Nairn and F. G. Stehli), pp. 325–420. Plenum Press, New York.

Siesser, W. G. (1978). Leg 40 results in relation to continental shelf and onshore geology. In *Initial Reports of the Deep Sea Drilling Project*, Vol. 40 (ed. H. M. Bolli *et al.*), pp. 965–79. US Government Printing Office, Washington, DC.

Silver, E. A. and Anderson, L. T. (1974). Possible left-lateral early to middle Mesozoic disruption of the southwestern North American craton margin (abst.). *Geological Society of America Abstracts with Program*, **6**, 955.

Sues, H.-D. and Taquet, P. (1979). A pachycephalosaurid dinosaur from Madagascar and a Laurasia–Gondwanaland connection in the Cretaceous. *Nature*, **279**, 633–5.

Supko, P. R., Perch-Nielsen, K., *et al.* (1977). Site 354. In *Initial Reports of the Deep Sea Drilling Projet*, Vol. 39 (ed. P. R. Supko *et al.*), pp. 45–100. US Government Printing Office, Washington, DC.

Taquet, P. (1978). Niger et Gondwana. *Ann. Soc. Géol. Nord*, 337–41.

Thiede, J. (1977). Subsidence of aseismic ridges: evidence from sediments on Rio Grande Rise (southwest Atlantic Ocean). *American Association of Petroleum Geologists Bulletin*, **61**, 939–40.

Thiede, J. (1979). History of the North Atlantic Ocean: evolution of an asymmetric zonal paleo-environment in a latitutinal ocean basin. In *Deep drilling results in the Atlantic Ocean: Continental margins and paleo-environment* (ed. M. Talwani, W. Hay, and W. B. F. Ryan), Maurice Ewing Series, Vol. 3, pp. 275–96.

Thierstein, J. R. (1979). Paleoceanographic implications of organic carbon and carbonate distribution in Mesozoic deep sea sediments. In *Deep drilling results in the Atlantic Ocean: Continental margins and paleoenvironment* (ed. M. Talwani, W. Hay, and W. B. F. Ryan), Maurice Ewing Series, Vol. 3, pp. 249–74.

Tomblin, J. F. (1975). The Lesser Antilles and Aves Ridge. In *The ocean basins and margins*, Vol. 3, The Gulf of Mexico and the Caribbean (ed. A. E. M. Nairn and F. G. Stehli), pp. 467–500. Plenum Press, New York.

Torquato, J. R. and Cordani, U. G. (1981). Brazil–Africa geological links. *Earth-Science Reviews*, **17**, 155–76.

Van der Voo, R., Mauk, F. J., and French, R. B. (1976). Permian-Triassic continental configuration and the origin of the Gulf of Mexico. *Geology*, **4**, 177–80.

Vogt, P. R. (1973). Early events in the opening of the North Atlantic. In *Implications of continental drift to the earth sciences*, Vol. 2 (ed. D. H. Tarling and S. K. Runcorn), pp. 693–712. Academic Press, London.

Walper, J. L. (1980). Tectonic evolution of the Gulf of Mexico. In *The origin of the Gulf of Mexico and the early opening of the central North Atlantic Ocean* (ed. R. H. Pilger), pp. 87–98. Proceedings of a symposium at Baton Rouge, Louisiana State University.

Wiedmann, J. and Neugebauer, J. (1978). Lower Cretaceous ammonites from the South Atlantic Leg 40 (DSDP), their stratigraphic value and sedimentological properties. In *Initial Reports of the Deep Sea Drilling Project*, Supplement to volumes 38, 39, 40, and 41, pp. 709–834. US Government Printing Office, Washington, DC.

3 PALAEOCLIMATIC HISTORY OF THE OPENING SOUTH ATLANTIC

Judith Totman Parrish

Climate has never been uniform in the South Atlantic Ocean. The rift system that opened to form the South Atlantic spanned nearly 60° latitude, and during its history the basin has extended across several climatic zones. Palaeoclimatic data for Gondwana as a whole are comparatively sparse, but for the circum-South Atlantic region, because the region has been studied intensively, palaeoclimatic history is fairly clear. In this chapter, methods that have been applied in studies of South Atlantic palaeoclimates will be discussed, followed by descriptions of what is currently known about the palaeoclimatic history of the South Atlantic and of southern Laurasia.

PALAEOCLIMATIC METHODS

A combination of two approaches to understanding palaeoclimates can be usefully applied to the South Atlantic. First, climatic patterns can be predicted using climatic models whose results can be compared with the available data. Second, inferences can be made about climate and climatic change from the distributions of organisms and of such climatically significant rocks as coals, evaporites, and laterites. A combination of the two approaches is advantageous since geological data for inferring palaeoclimates are scarce for most geological ages and can be used to delimit only relatively local climatic patterns. Theoretical climate models can provide a larger context in which to interpret the geological data.

Climate models

Conceptual and numerical climate models have been published for the middle part of the Creta-ceous (Gordon 1973; Parrish and Curtis 1982; Parrish *et al.* 1982; Barron and Washington 1982, 1984; Lloyd 1982); and all show that palaeogeography is a major agent of change in atmospheric circulation patterns. The patterns that are predicted depend on the positions of the continents and how the continents would be expected to have modified the zonal circulation is, which is the most fundamental component of atmospheric circulation. This zonal circulation is associated with low barometric pressure and high rainfall at the Equator and at about 60° north and south, and low barometric pressure and low rainfall at the poles and at about 30° north and south. These climatic zones move 10–15° north and south with the seasons.

Even during times of maximum zonality, east–west climatic asymmetry across continents exists because of differences in the source and nature of the winds encountering the coasts. For example, rainfall in South Carolina is higher than in central California, which is at the same latitude. The prevailing winds in South Carolina are from the south, away from the Equator, and are warm and moist, whereas the prevailing winds in central California are from the north, away from the poles, and are cool and carry relatively less moisture. In general, east coasts are wetter than west coasts between about 40° north and south and drier poleward of 40° north and south (Köppen, in Robinson 1973).

The distribution of rainfall also depends on the size of the continents, on the height, breadth, and location of mountain ranges, and on the sources of moisture for winds flowing over the continents. In general, a large continent will have a dry interior because winds are depleted of moisture as they encounter coastlines and mountains. Even a low-presssure cell (ascending air) will not generate rain if the air flowing into the cell comes mainly from continental regions. Mountains will have a variety of effects, depending on their geography. A coastal mountain range may create a rain shadow, making the continental interior dry. Alternatively, an interior mountain range will tend to make the continental interior wetter by causing the release of moisture; mountain slopes and uplands can be much wetter than the regions surrounding them. Therefore, although the zonal pattern of rainfall is a good first approximation for predicting palaeoclimatic patterns, this pattern is strongly modified by various palaeogeographical situations (Gyllenhaal *et al.* 1991; Patzkowsky *et al.* 1991).

Evaporation plays an equally important role to that of precipitation in determining overall climate. The maps of Parrish *et al.* (1982) showed only rainfall, but their analysis of coal and evaporite data revealed the importance of evaporation patterns, which are impossible to predict conceptually except in the most general way. Evaporation is related to temperature, which decreases away from the Equator. The result is that, although precipitation in the high-latitude humid zones is relatively lower than along the Equator, those zones are nevertheless likely to be wet because evaporation is also lower.

The rainfall predictions of Parrish *et al.* (1982), Barron and Washington (1982), and Barron (1990) are the most relevant to the discussions in this book. Parrish *et al.* (1982) assumed stabilty of Hadley cell circulation (in which air rises at the equator, and descends in the subtropics) and a dominant influence of palaeogeography on global climate (Parrish 1982). The predictions thus included an equatorial wet belt as this belt is a dominant feature of the general circulation of the atmosphere today. The dispersion of continental land masses in the middle Cretaceous would not have greatly disrupted zonal atmospheric circulation, as did Pangaea earlier in the Mesozoic (Parrish and Doyle 1984; J. M. Parrish *et al.* 1986). This equatorial wet belt would have fluctuated perhaps 10° in latitude with the seasons.

The results of Barron and Washington's (1982) and Barron's (1990) numerical model runs are similar in some ways to the predictions derived from the conceptual models of Parrish *et al.* (1982), particularly in the conclusion that climate would have been highly zonal in the middle Cretaceous. In Barron and Washington's (1982) simulations, the equatorial rain belt fluctuated between about 10° south of the Equator in the northern winter and 20° north (over Tethys) in the northern summer; the equatorial areas of northern Africa and South America were dry in their simulations. However, in Barron's (1990) more recent simulations, the peak precipitation zones migrate between about 5° north and 5° south. Barron (1990) did not present an annual precipitation map, but the equatorial rain belt would have been similar to that of Parrish *et al.* (1982). Parrish *et al.* (1982) and Barron and Washington (1982) agreed on the importance of Tethys for determining the position of the northern subtropical high-pressure belt and the consequent relatively high rainfall in eastern North America and south-western Europe.

Geological data on terrestrial climates

Several types of rocks are constrained by climate for their formation and, therefore, their occurrence can be indicative of certain climatic conditions. Among the most useful rocks for interpreting terrestrial palaeoclimates are coals, which indicate relatively wet conditions (high precipitation, low evaporation, or both; Gyllenhaal *et al.* in press); evaporites (salt deposits) and aeolian sandstones (ancient sand dunes), which indicate dry conditions (Borchert and Muir 1964; McKee 1979); and laterites and bauxites (red, iron- and aluminium-rich tropical soils), which indicate warm and wet conditions (Nicolas and Bildgen 1979). Red beds ('savannah red beds' or 'desert red beds') are regarded as indicative of a generally warm climate with a strongly seasonal wet–dry cycle (Van Houten 1982). Marine rocks can yield palaeoclimatic information that may be useful for understanding the climate of at least the coastal zones. These include certain clay minerals, which some workers believe reflect onshore climatic conditions (Millot 1970; Chamley 1979) and oolitic ironstones, which form in close near-shore marine environments adjacent to regions that are undergoing lateritic weathering (Van Houten and Bhattacharya 1982). In addition, terrestrial organic material and pollen grains, which may provide clues to the types of plants in and the productivity of terrestrial biological communities, commonly occur in marine sections. All these types of information must be used carefully. A single coal deposit, for example, may reflect climate only indirectly, if it was formed in an area that had a low rainfall but a high water table owing to long-distance groundwater flow or a high base level. A pattern formed by many coal deposits, however, may be indicative of high rainfall (McCabe and Parrish in press; Gyllenhaal et al. 1991).

PALAEOCLIMATIC HISTORY OF THE OPENING SOUTH ATLANTIC

Climate in the vicinity of the opening South Atlantic would have been influenced not only by the opening of the basin itself but also by the climatic effects of palaeogeographical changes extrinsic to the basin. In the following pages, consideration will therefore be given to global climatic patterns, with particular reference to Gondwana and southern Laurasia, in order to provide a context for the local climatic changes observable in the rock record. Maps of postulated rainfall for the Barremian (Early Cretaceous; Parrish 1985), Cenomanian (early Late Cretaceous), Maastrichtian (late Late Cretaceous) and Lutetian (Eocene; Parrish et al. 1982) are reproduced in Figures 3.1 and 3.3–5, along with climatically significant sediment data for the Early Cretaceous (Neocomian–Barremian, Fig. 3.1), Aptian–Albian (Fig. 3.2), Late Cretaceous (Cenomanian–Campanian, Fig. 3.3), Maastrictian (Fig. 3.4), Eocene (Fig. 3.5), and Oligocene (Fig. 3.6). Most data on climatically

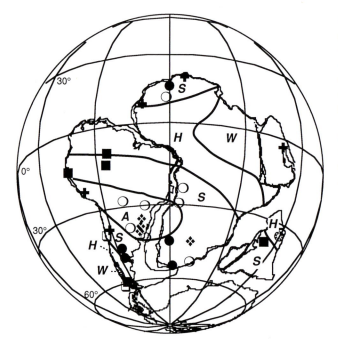

Fig. 3.1 Palaeogeography and predicted precipitation in the Barremian (middle Early Cretaceous; Parrish 1985), and sedimentological palaeoclimatic indicators for the Neocomian through Barremian. ⊡, Shelf carbonate and reefs; ■, coal; ▲, bauxite; ▼, laterite; ◆, kaolinite; ●, red beds; +, evaporite; ✦, aeolian sandstone; ◐, ironstone; ○, 'continental carbonate deposits with gypsum' (Ronov and Balukhovskii 1982); *A*, arid; *S*, semi-arid/semi-humid; *H*, humid; *W*, wet; ——, isohyets. See text for data sources.

Fig. 3.2 Continental positions (Denham and Scotese 1988) and palaeoclimatic indicators for the Aptian–Albian (late Early Cretaceous). ☶, Aptian–Albian salts of the northern South Atlantic; ✳, pollen of Araucariaceae (Doyle *et al*. 1982).

Fig. 3.3 Palaeogeography (Ziegler *et al.* 1983) and predicted precipitation (modified slightly from Parrish *et al.* 1982) for the Cenomanian (early Late Cretaceous), and sedimentological palaeoclimatic indicators for the Cenomanian through Campanian. ᵒᵒᵒ, Cenomanian shoreline in northern Africa. Additional symbols as in Fig. 3.1. See text for data sources.

Fig. 3.4 Palaeogeography (Ziegler *et al.* 1983), predicted precipitation (modified slightly from Parrish *et al.* 1982), and sedimentological palaeoclimatic indicators for the Maastrichtian (latest Cretaceous). Symbols as in Fig. 3.1. See text for data sources.

Fig. 3.5 Palaeogeography (Ziegler *et al.* 1983), predicted precipitation (Parrish *et al.* 1982), and sedimentological palaeoclimatic indicators for the Eocene. ▒, Areas of extensive bauxitization. Additional symbols as in Fig. 3.1. See text for data sources.

Fig. 3.6 Continental positions (Denham and Scotese 1988) and sedimentological palaeoclimatic indicators for the Oligocene. Symbols as in Fig. 3.1. See text for data sources.

significant rocks (i.e., evaporites, laterites, coals, etc.) are taken from sources listed in Parrish et al. (1982) and from Ronov and Balukhovskii (1982); those papers should be referred to where specific citations are not noted. Additional data sources are referred to individually.

In the figures, 'arid' is roughly equivalent to desert climate ('low rainfall' in the following discussion, or less than 10 inches of rain per year), 'semi-arid/semi-humid' is roughly equivalent to steppe, taiga, and Mediterranean ('moderately low rainfall', 10–30 inches), 'humid' is roughly equivalent to humid subtropical or maritime climates ('moderately high rainfall', 30–60 inches), and 'wet' is roughly equivalent to tropical rainforest, coastal rainforest, monsoonal, or tropical savannah climates ('high rainfall', more than 60 inches). Seasonality within these categories can be great or small. For example, Kwang-Tri, Vietnam, which lies within the influence of the Asian monsoon, receives 86 per cent of its annual rainfall in the six wettest continguous months. By contrast, Belem, Brazil, in the Amazon Basin, receives only 73 per cent and Evangelists' Island, Chile, only 53 per cent. All these stations receive between 100 and 110 inches rainfall annually and all lie at elevations lower than 50 feet.

Late Jurassic

Late Jurassic climatic patterns reflect the vestiges of the monsoonal circulation that dominated global climate in the Permian and Triassic, when Pangaea was fully assembled (Robinson 1973; Rowley et al. 1985; J. M. Parrish et al. 1986). The effects of monsoonal climate, namely highly seasonal rainfall in the equatorial region and on the east side of Pangaea, are strongly reflected in deposits of the earlier Mesozoic, and were still in evidence in the Late Jurassic. Evaporites are abundant near the palaeoequator in northern Africa, the Middle East, and southern North America (Parrish et al. 1982). A few coal deposits of Late Jurassic age are, however, found near the evaporites in northern Africa and southern

Mexico, possibly indicating that rainfall was seasonal. Robinson (1973) invoked monsoonal seasonality to explain a similar geographical mixture of coals and evaporites in the Triassic. Small changes in temperature could alter the predominance of dry- or wet-season indicators in the geological record (Parrish in press).

Late Jurassic coal deposits are found in a belt from southern South America to north-eastern Australia, in the predicted wet belt rimming southern Pangaea (Parrish *et al.* 1982). Araucariaceae, which are considered indicative of tropical, wet climates (Doyle *et al.* 1982), and frogs (Bonaparte 1978) occur in ?Late Jurassic rocks of southern America. Evaporites and aeolian sandstones are found at subtropical latitudes on the western side of Gondwana (Volkheimer 1967; Bonaparte 1978; Petri and Campanha 1981).

Rainfall would have been low in the continental interior. The region immediately bordering the opening South Atlantic was near the centre of the continent and is thus predicted to have been arid to semi-arid. However, the altitude of the rift uplift would probably have prevented the region from being very dry, either because the uplifted region would have extracted whatever moisture was in the air or because the cooler temperatures encountered at high altitudes would have resulted in lower evaporation rates. North of the rift uplift, in the region of the future western Gulf of Guinea, rainfall is thought to have been moderate, increasing toward the palaeoequator. Climate south of the rift uplift would have depended on the height of the Late Jurassic Andes—the higher the mountains, the drier the continental interior would have been. Because the mountain chain, as reconstructed by Ziegler *et al.* (1983), was not continuous, Parrish and Curtis (1982) predicted low to moderately high rainfall for southern South America, South Africa, and the Falkland Plateau during the Late Jurassic.

Palaeoclimatically significant deposits are uncommon in Late Jurassic rocks from the interior of Gondwana. Aeolian sandstones of Triassic and/or Jurassic age are found in the southern part of the rift zones, in the vicinity of present-day South-west Africa (Martin 1973) and in Argentina (Volkheimer 1967). Late Jurassic evaporites occur just north of the rift uplift, in what is now easternmost Brazil. The evaporites are overlain by deposits containing large conifer trunks and floodplain and lacustrine deposits (Petri and Campanha 1981); the transition may reflect the drift of the region towards the Equator into the zone of higher rainfall.

Early Cretaceous (Fig. 3.1)

Gondwana's position at low latitudes and the opening seaways between Antarctica–Australia, India, and South America–Africa would have weakened the monsoonal circulation during the Early Cretaceous. The coal belt along the southern margin of Pangaea persisted through the Early Cretaceous, as did the evaporite belt around the northern, western, and north-eastern margins, indicating that monsoonal conditions probably continued until the South Atlantic was completely open. Figure 3.1 shows palaeoclimatic indicators for the Neocomian through Barremian, and palaeogeography and rainfall predictions for the Barremian (Parrish 1985).

Neocomian

In the circum-South Atlantic region, floodplain and lacustrine deposits persisted through the Neocomian in easternmost Brazil (Petri and Campanha 1981) and in Gabon and Congo (Doyle *et al.* 1982). Petri and Campanha (1981) regarded the floodplain deposits, which consist of red beds, as representing a more humid, but still seasonally dry, climate. Similarly, Doyle *et al.* (1982) suggested that the climate recorded in the African lake deposits was semi-arid or wetter, but with a seasonal wet–dry cycle. Red beds extended southward to northern Patagonia (Volkheimer 1967; Ronov and Balukhovskii 1982). Red beds and alluvial sediments suggestive of a semi-arid climate occur at the same palaeolatitude in southern Africa (Tankard *et al.* 1982; Ronov and Balukhovskii 1982; Karpeta 1987). Throughout the Early Cretaceous, evaporites and aeolian sandstones indicate an arid region to the north and east, in what is now northern Chile and southernmost Brazil (Volkheimer 1967; Hallam 1984; Riccardi 1988); Ronov and Balukhovskii (1982, Fig. 3.1B) have illustrated aeolian sandstones in northern South Africa, although no additional information was provided. Neocomian coals occur in southern Patagonia (Ronov and Balukhovskii 1982; Riccardi 1988), and Ronov and Balukhovskii 1982) have illustrated coals/lignites in the region just south of the Equator.

Thus, a well-defined zonation of climate with respect to precipitation–evaporation can be documented for the southern part of the South Atlantic region in the Neocomian, consisting of a dry belt extending from about 30°S to 45°S palaeolatitude (Volkheimer 1967; Hallam 1984; Riccardi 1988), a semi-arid to semi-humid zone with seasonal rainfall between about 45°S and 55°S palaeolatitude (Volkheimer 1967; Dingle *et al.* 1973; McLachlan and McMillan 1976; Tankard *et al.* 1982; Karpeta 1987; Riccardi 1988), and a humid zone south of 55°S palaeolatitude (Volkheimer 1967; Riccardi 1988). The belt of seasonal rainfall contains indicators of at least seasonally dry climate (desiccation features, gypsum and pseudomorphs after gypsum rosettes, red beds; Dingle *et al.* 1973; Karpeta 1987), but also abundant plant material and large fossilized logs (Volkheimer 1967; McLachlan and McMillan 1976; Tankard *et al.* 1982). Overall, the climate appears to have been warm. This is inferred from the presence of limestones in central Chile (about 40°S palaeolatitude; Volkheimer 1967), coral patch reefs in southern Chile (about 60°S palaeolatitude), corals in South Africa (about 55°S palaeolatitude; Karpeta 1987), and warm-water dinoflagellates in southern Argentina (about 50°S palaeolatitude; Quattrocchio 1982).

Barremian (*Fig. 3.1*)

By Barremian time, Gondwana was partly broken up; seaways existed between India, Africa, and Antarctica–Australia and the southern South Atlantic was partly open. The initial effect of the breakup on the region of the rift would have been to make the region somewhat drier, due to the altitudinal collapse of the rift zone within the rift uplift (Hay *et al.* 1982; Doyle *et al.* 1982) and its position in low latitudes. The floodplain and lacustrine deposits in eastern Brazil (Petri and Campanha 1981) gave way in the Barremian to fluvial and eolian deposits (Bigarella 1973; Petri and Campanha 1981; Riccardi 1988) and evaporitic deposition persisted in northern Chile south of about 25°S palaeolatitude (Volkheimer 1967; Riccardi 1988). Fluvial and lacustrine deposition persisted into the early Aptian in Gabon (Doyle *et al.* 1982), but geochemical data (F. Walgenwitz, cited in Doyle *et al.*, 1982) have suggested that the lakes were more saline than earlier in the Cretaceous. Otherwise, climatic zonation was similar to that in the Neocomian (Volkheimer 1967; Dingle *et al.* 1973; Riccardi 1988), with plant-rich red beds occupying a zone about 45°S to 55°S between an arid zone to the north and a humid zone to the south (Volkheimer 1967, 1980; Dingle *et al.* 1973; Doyle *et al.* 1982; Archangelsky and Seiler 1980; Quattrocchio 1982). Brenner (1976) placed the boundary between the Northern and Southern Gondwana palynoprovinces at approximately 30°S palaeolatitude, but this may have been a generalization based on the relative lack of data between the northern South Atlantic and southern Africa and South America. The Southern Gondwana province is represented in the plant-rich red beds in northern Patagonia and South Africa.

Aptian–Albian (*Fig. 3.2*)

Thick Aptian–Albian evaporites are found on both sides of the northern South Atlantic in the Aptian and Albian (Franks and Nairn 1973; Asmus and Ponte 1973; de Klasz 1978), and the climate is generally considered to have been very arid (de Klasz 1978; Arthur and Natland 1979) in the rift. Where the marine waters in the South Atlantic basin were fresh enough, reefs and oolites were deposited in Africa and South America (Franks and Nairn 1973; Falkenhein *et al.* 1981). The climate was apparently arid to semi-arid in the continental regions flanking the rift. Ronov and Balukhovskii (1982, Figs. 1C and 2A) indicated the presence of abundant 'continental carbonate deposits with gypsum' and 'carbonate-bearing red beds and variegated beds' in those regions. Continental and arid lacustrine sediments predominate in central Argentina (Bonaparte 1978; Riccardi 1988).

The palynofloras of the Northern Gondwana province (Brenner 1976), characterized by an abundance of *Classopollis*, are generally consistent with a semi-arid, monsoonal climate (Doyle *et al.* 1982, and references therein). *Classopollis* dominates the palynofloras from the Cape and Angola Basins (Dingle *et al.* 1973; McLachlan and Pieterse 1978; Morgan 1978) and from Aptian–Albian rocks in the northern South Atlantic (Doyle *et al.* 1977; Doyle *et al.* 1982). However, a narrow wet belt along the palaeoequator in north-western Africa (Senegal and the Gulf of Guinea region) and lying outside the rift zones is suggested by the abundance of pollen of the Araucariaceae (Doyle *et al.* 1982; Fig. 3.2). In addition, the northernmost rift basins contain gypsum rather than halite, which suggests that the water was slightly fresher than farther south, possibly owing to increased runoff (Doyle *et al.* 1982). According to Ronov and Balukhovskii (1982, Fig. 2A), kaolinite, commonly a product of intense chemical weathering and implying wet conditions, occurs in the region of Cameroon in the Albian.

The climate in south-western South America has been described as humid (Volkheimer 1967), although the presence of red beds suggests that the rainfall was seasonal. The flora has an abundance of podocarp pollen (Doyle *et al.* 1982) and fern spores (Archangelsky and Seiler 1980), which support the suggestion that the climate was wetter than that farther north (Doyle *et al.* 1982). Ronov and Balukhovskii (1982, Figs 1C and 2A) illustrate the presence of coal/lignite, humid-climate floras, and kaolinite in southern South Africa for the Aptian–Albian interval.

Late Cretaceous

Cenomanian–Campanian (*Fig. 3.3*)

In Fig. 3.3, data are for the Cenomanian through Campanian and palaeogeography and rainfall predictions for the Cenomanian (Parrish *et al.* 1982; Ziegler *et al.* 1983). The northern African shoreline for the Cenomanian is shown as it was far from the continental margin, and the isohyets, which are for the continental regions, end at the shore line.

By the Cenomanian (earliest Late Cretaceous), the South Atlantic was completely open and no vestiges of monsoonal climate would have remained. Moreover, biotic exchanges among the Gondwanan continents would have been inhibited more by geography than by climate (see Chapter 2). North-western Africa was far enough north by Cenomanian time to have penetrated the dry subtropics, as indicated by evaporites in Algeria and Libya (Parrish *et al.* 1982); the continent continued to move north, bringing the dry belt southward across the continent. Lignite and swamp vegetation of Cenomanian age are found in northeastern Brazil (Petri and Campanha 1981). Late Cretaceous coal, ironstone, kaolinite, and bauxite also occur in Africa and South America just south of the palaeoequator (Ronov and Balukhovskii 1982). These deposits suggest a wet belt centred about 15°S, coincident with the location of the rain belt predicted in Barron and Washington's (1982) simulations for the southern summer and with the annual rainfall belt proposed by Parrish *et al.* (1982).

Ronov and Balukhovskii's (1982) and Hallam's (1984) data for the Late Cretaceous show that evaporites persisted in northern Africa and in the South Atlantic where the ocean was narrowest, especially in eastern Brazil (Bonaparte 1978). That the climate was warm in that region is suggested by widespread shelf carbonates and reefs (Asmus and Ponte 1973; Mabesoone *et al.* 1981; Ronov and Balukhovskii 1982); in Africa, the Turonian Wadatta Limestone of Nigeria shows evidence of phreatic diagenesis (Nwajide 1986), which is more consistent with the wet-climate indicators on that side of the ocean. The region lying at the boundary between the North African dry belt and the wet belt to the south is characterized by red beds (Reyment 1981; Ronov and Balukhovskii 1982);

Evaporites, red beds, and aeolian sandstones of Late Cretaceous age occur throughout most of central South America from about 10°S to 55°S palaeolatitude in western and central South America (Volkheimer 1967; Riccardi 1988). A 'tropical to subtropical' humid zone, expressed in the floras, paralleled the arid one to the west south of 35°S (Volkheimer 1967). These belts probably reflect the presence of and were controlled by the Andean arc (Parrish *et al.* 1982; Riccardi 1988). In contrast to Africa, South America did not move much latitudinally during the Late Cretaceous.

No evidence exists for the humid region predicted by Parrish *et al.* (1982) for south-eastern Brazil. This humid region was based on a low-reliability prediction of the development of the South Atlantic subtropical high-pressure cell (Parrish *et al.* 1982). The point in the history of an ocean basin at which it begins to create its own climate system (for example subtropical high-pressure cells; Parrish *et al.* 1983), rather than being influenced largely by the adjacent continents, is not a problem that can be adequately addressed by conceptual models such as those of Parrish *et al.* (1982), so the rainfall prediction is hereby modified accordingly. Similarly, the south Tethyan subtropical high-pressure cell apparently did not affect eastern Africa and India as much as that predicted by Parrish *et al.* (1982). For both these cases, however, the discrepancy between model and data may be a fault of the model only in so far as the model predicts rainfall but not evaporation. As the Late Cretaceous was an extremely warm time, the lack of humid-climate indicators in these low-latitude regions may be more an expression of higher evaporation rates than lower rainfall.

Maastrichtian (Fig. 3.4)

Climatic zones across Africa and South America migrated south with the northward drift of the continents. Maastrichtian (latest Cretaceous) rainfall patterns in the circum-South Atlantic region are thought to have been very similar to those in the Cenomanian, the major difference being the generally wetter equatorial regions permitted by the width of the Central Atlantic, reflected in laterite, coal, kaolinite, and ironstone deposits of Central Africa (Ronov and Balukhovskii 1982; Parrish *et al.* 1982). Note that the rising Zagros Mountains apparently created a rainshadow in the path of the equatorial easterlies for the Arabian Peninsula. In southern South America, the climate remained arid behind the Andean arc and across the continent at about 30°S (the predicted dry belt) and was more humid farther south (Riccardi 1988). The climate in the dry belt was, however, at least seasonally wet, as suggested by the presence of red beds (Ronov and Balukhovskii (1982) and turtles and crocodilians (Bonaparte 1978). Vegetation also indicates humid climate in South Africa (Tankard *et al.* 1982), but the climate in the southern portion of the continent was probably more zonal than indicated by Parrish *et al.* (1982), as evaporites were deposited in the border region between South Africa and Mozambique (Dingle *et al.* 1973); the rainfall map is modified in accordance with the data. As in the Cenomanian (Fig. 3.3), the humid belt in south-eastern Africa depended on the influence of the winds in the eastern limb of the southern Tethyan subtropical high-pressure cell (Parrish *et al.* 1982). However, the Maastrichtian was cooler than the earlier part of the Late Cretaceous, so the absence of humid-climate indicators is not as likely to be a result of higher evaporation (see next paragraph). These winds may have been blocked or moved too far east by India to affect south-eastern Africa as strongly as predicted by Parrish *et al.* (1982).

The data suggest that the equatorial humid belt may have become wider in the Maastrichtian than earlier in the Cretaceous, although the degree of scatter in the data does not permit a confident conclusion that this is not merely an artefact of the distribution of the data. An apparent expansion of the humid zone does not necessarily imply a change in the distribution of rainfall. Global cooling would also bring about an expansion of humid zones through reduction of evaporation potential. Basing his arguments largely on conclusions made by Bonaparte (1978) about the distribution of vertebrates, Riccardi (1988) suggested that low-latitude temperatures in South America were lower in the Maastrichtian than earlier in the Cretaceous. Cooler temperatures might be sufficient to result in a wider distribution of wet-climate indicators, without invoking changes in rainfall, by decreasing the evaporation potential (Parrish et al. 1982; Parrish in press). Isotopic (Savin 1977) and floral (Wolfe 1980) evidence both indicate that climate in the Maastrichtian probably was cooler than earlier in the Cretaceous.

Early Tertiary

Palaeocene and Eocene

Data presented by Ronov and Balukhovskii (1982) indicate that Palaeocene climates in the circum-South Atlantic region were not greatly different from those of the Maastrichtian. The major difference was that the wet belt just south of the equator may have been slightly narrower. In the Eocene (Fig. 3.5), climate patterns do not appear to have been very different, but the effect of climate on sedimentation was dramatically different. Eocene bauxites are very abundant in central Africa and northern South America together with ironstone and kaolinite deposits and coals (Ronov and Balukhovskii 1982; Prasad 1983). Prasad (1983) referred to this as the 'early Tertiary bauxite event', which was also expressed in southeastern Africa, Madagascar, and India. Prasad (1983) suggested that the deposits represented a humid phase that peaked in the Eocene. However, the formation of bauxites and related deposits is also enhanced by warmth, and the Eocene is widely recognized as the warmest interval in the Tertiary (Savin 1977; Estes and Hutchison 1980; Wolfe 1980; Shackleton and

Boersma 1981; Murphy and Kennett 1985). Warmth and, indirectly, high humidity are supported by the fauna of the Ameki Formation of Nigeria, which is interpreted as having lived in 'tropical' waters in which salinity fluctuated from normal marine to brackish (Arua 1988). Clay minerals in the south-eastern South Atlantic are suggestive of intense lateritic weathering, that is, high temperatures and abundant rainfall (Nicolas and Bildgen 1979) on the adjacent African continent, particularly north of 30°S (Robert 1980; Chamley 1986). In South-west Africa, the late Palaeocene–early Eocene Bundtfelshuh beds comprise aeolian sandstone and calcrete (Dingle et al. 1973). In South America, the Andes continued to exert an influence, creating the predicted rain shadow in north-eastern Argentina (Volkheimer 1967; Parrish et al. 1982; Riccardi 1988). Further south, however, the climate was wetter, as suggested by the coals. The increased humidity may have been partly due to the cooler temperatures, as the floras and vertebrates are consistent with a cool-temperature climate (Volkheimer 1971).

Oligocene (Fig. 3.6)

By the Oligocene, the south polar front was identifiable in the South Pacific and South Atlantic (Murphy and Kennett 1985; Wise et al. 1985). Clay minerals in deep sea drilling cores are consistent with a cooler climate (Robert 1980; Chamley 1986), and both evaporites and bauxites, which partly depend on elevated temperatures for their formation, were much less widespread (Ronov and Balukhovskii 1982). An Oligocene freshwater fish fauna in southern Argentina contains no tropical forms; the climate suggested was warm temperate ('temperature not very high'; Cione and Exposito 1980). Floras in South Africa were described as temperate rainforest (Tankard et al. 1982).

BRIEF CLIMATIC HISTORY OF LAURASIA

Late Jurassic

The monsoonal climate regime was disrupted first in the northern hemisphere with the opening of Tethys, remaining in a much-reduced form in central Asia (Mongolia and northern China). Tethys was a continuous, albeit narrow, east–west seaway by Late Jurassic time, contributing to the breakup of the monsoon and the establishment of normal zonal dry conditions in the north subtropics, along the southern margin of Laurasia. Evaporite deposits of latest Jurassic age are abundant all along the southern margin of Laurasia, from what is now western China to Texas (Cook and Bally 1975; Parrish *et al.* 1982). Abundant coals are found in high-latitude portions of western North America and eastern Laurasia.

Early Cretaceous

By the middle Early Cretaceous, Tethys was probably wide enough to generate its own subtropical high-pressure cell during the northern winter. The westerlies generated in this cell would have supplied moisture to eastern North America and south-western Europe during the winter. Progressive stages in the opening of the North Atlantic and the consequent strengthening of the high-pressure cell are reflected very well in the progressive geographical replacement of evaporites by coals and bauxites, starting in eastern North America and south-western Europe in Neocomian time and extending well into the Russian platform by Aptian–Albian time (Nicolas and Bildgen 1979). Aptian–Albian palynofloras of southern Laurasia also indicate humid conditions (Brenner 1976). Seasonality of the rainfall may be indicated by the persistence of scattered evaporites in south-western Europe. In addition, Chamley (1979) interprets changes in clay mineralogy from the North Atlantic as indicating a change from dry to seasonally wet conditions during the Early Cretaceous.

Late Cretaceous

Cenomanian rainfall patterns in Laurasia would have been very similar to Barremian ones. However, the Tethyan subtropical high pressure cell is predicted to have been year-round by the Cenomanian, carrying rainfall year-round to eastern North America and south-western Europe. Indeed, evaporites are unknown from south-central Laurasian rocks of Cenomanian age, while bauxites are common in middle Cretaceous rocks of Apulia and south-western Europe and Cenomanian coals occur all along the southern margin of Laurasia. Parrish *et al.* (1982) predicted that eastern Europe might have been slightly drier in the Cenomanian than Parrish and Doyle (1984) predicted for the Barremian, owing to the influence of the mountain front that now comprises the Tien Shan and Kunlung Mountains, the Hindu Kush and Pamirs, and the Elburz and Caucasus Mountains. Evaporites do occur just behind the mountain front in what was the southern USSR, but coals formed farther north, probably reflect the year-round presence of the subtropical high-pressure cell, which would have tended to make the region more humid.

Early Tertiary

The general pattern established in the early Late Cretaceous continued into the Tertiary. South–central Laurasia remained humid, as indicated by the presence of numerous coal deposits, laterites, and bauxites of Palaeocene through Oligocene age. Palaeocene and Eocene evaporites are common in the subtropical regions of eastern and western Laurasia, in what is now China and the southern part of North America, and a few evaporite deposits of Eocene age occur in Spain, near the developing eastern Atlantic upwelling zone.

CONCLUSIONS

Climate could have affected organismal exchange between Africa and South America in the late

Early Cretaceous. The narrowest point in the opening South Atlantic was at the northern end of a vast evaporite basin, and the presence of thick and extensive evaporites suggests that the climate there was hot and dry. Thus, organisms that were adapted to the relatively cool, humid temperate climates of southern South America would have been prevented from migrating between the continents after the middle of the Early Cretaceous. After the Aptian–Albian, faunal exchange would have been inhibited more by the barrier created by the opening South Atlantic.

REFERENCES

Archangelsky, S. and Seiler, J. (1980). Algunos resultados palinológicos de la Perforación UN OIL OS-1, del so de la Provincia de Chubut. II Congreso Argentino de Paleontología y Bioestratigrafía y 1 Congreso Latinoamericano de Paleontología, 2-6 April 1978, *Actas*, Vol. II, Buenos Aires, pp. 215–28.

Arthur, M. A. and Natland, J. H. (1979). Carbonaceous sediments in the North and South Atlantic: the role of salinity in stable stratification of Early Cretaceous basins. In *Deep drilling results in the Atlantic Ocean: Continental margins and paleoenvironment* (ed. M. Talwani, W. Hay, and W. B. F. Ryan), *Maurice Ewing Series*, Vol. 3, pp. 375–401. American Geophysical Union, Washington DC.

Arua, I. (1988). Paleoecology of the Eocene Ameki Formation of southeastern Nigeria. *Journal of African Earth Sciences*, **7**, 925–32.

Asmus, H. E. and Ponte, F. C. (1973). The Brazilian marginal basins. In *The ocean basins and margins* (ed. A. E. M. Nairn and F. G. Stehli), Vol. 1, The South Atlantic, pp. 87–133. Plenum Press, New York.

Barron, E. J. (1990). *Atlas of Cretaceous model results*. Earth System Science Center Publication, The Pennsylvania State University, University Park.

Barron, E. J. and Washington, W. M. (1982). Cretaceous climate: a comparison of atmospheric simulations with the geologic record. *Palaeogeography, Palaeoclimatology, Palaeoecology*, **40**, 103-33.

Barron, E. J. and Washington, W. M. (1984). The role of geographic variables in explaining paleoclimates: results from Cretaceous climate model sensitivity studies. *Journal of Geophysical Research*, **89**, 1267–79.

Bigarella, J. J. (1973). Geology of the Amazon and Parnaiba Basins. In *The ocean basins and margins* (ed. A. E. M. Nairn and F. G. Stehli), Vol. 1. The South Atlantic, pp. 25–86. Plenum Press, New York.

Bonaparte, J. F. (1978). *El Mesozoico de America del Sur y sus Tetrápodos*. Opera Lilloana, Vol. 26. Ministerio de Cultura y Educación, Tucuman, Argentina.

Borchert, H. and Muir, R. (1964). *Salt deposits—the origin, metamorphism, and deformation of evaporites*. D. Van Nostrand Co., London.

Brenner, G. J. (1976). Middle Cretaceous floral provinces and early migrations of angiosperms. In *Origin and early evolution of angiosperms* (ed. C. B. Beck), pp. 23–47. Columbia University Press, New York.

Chamley, H. (1979). North Atlantic clay sedimentation and paleoenvironment since the Late Jurassic. *Deep drilling results in the Atlantic Ocean: Continental margins and paleoenvironment* (ed. M. Talwani, W. Hay, and W. B. F. Ryan), *Maurice Ewing Series*, **3**, 342–61.

Chamley, J. (1986). Clay mineralogy at the Eocene/Oligocene boundary. In *Terminal Eocene events* (ed. Ch. Pomerol and I. Premoli-Silva), pp. 381–6. Elsevier, Amsterdam.

Cione, A. L. and Expósito, E. (1980). Chondrichthyes (Pisces) del 'Patagoniano' S.L. de Astra, Golfo de San Jorge, Prov. de Chubut, Argentina. Su significación paleoclimática y paleobiogeográfica. II. Congreso Argentino de Paleontología y Bioestratigrafía y I Congreso Latinoamericano de Paleontología, 2-6 April 1978, Buenos Aires, *Actas*, **II**, 275–90.

Cook, T. D. and Bally, A. W. (1975). *Stratigraphic atlas of North and Central America*. Princeton University Press, Princeton.

de Klasz, I. (1978). The West African sedimentary basins. In *The Phanerozoic geology of the world* (ed. M. Moullade and A. E. M. Nairn), Vol. II, The Mesozoic A, pp. 371–99. Elsevier, Amsterdam.

Denham, C. R. and Scotese, C. R. (1988). *Terra Mobilis*©: A Plate Tectonic Program for the Macintosh®. Published privately.

Dingle, R. V., Siesser, W. G., and Newton, A. R. (1973). *Mesozoic and Tertiary geology of Southern Africa*. AA. Balkema, Rotterdam.

Doyle, J. A., Biens, P., Doerenkamp, A., and Jardiné, S. (1977). Angiosperm pollen form the pre-Albian Lower Cretaceous of equatorial Africa. *Bulletin Centres Recherche Exploration–Production Elf Aquitaine*, **1**, 451–73.

Doyle, J. A., Jardiné, S., and Doerenkamp, A. (1982).

Afropollis, a new genus of early angiosperm pollen, with notes on the Cretaceous palynostratigraphy and paleonenvironments of Northern Gondwana. *Bulletin Centres Recherche Exploration–Production Elf Aquitaine*, **6**(1), 39–117.

Estes, R. and Hutchison, J. H. (1980). Eocene Lower vertebrates from Ellesmere Island, Canadian Arctic Archipelago. *Palaeogeography, Palaeoclimatology, Palaeoecology*, **30**, 325–47.

Falkenhein, F. U. H., Franke, M. R., and Carozzi, A. V. (1981). Petroleum geology of the Macaé Formation (Albian–Cenomanian), Campos Basin, Brazil. Ciência, Técnia, Petróleo Seccão: Exploracáo de Petróleo, **11**, 140 pp.

Franks, S. and Nairn, A. E. M. (1973). The equatorial marginal basins of west Africa. In *The ocean basins and margins* (ed. A. E. M. Nairn and F. G. Stehli), Vol. 1, The South Atlantic, pp. 301–50. Plenum Press, New York.

Gordon, W. A. (1973). Marine life and ocean surface currents in the Cretaceous. *Journal of Geology*, **81**, 269–84.

Gyllenhaal, E. D., Engberts, C. J., Markwick, P.J., Smith, L. H., and Patzkowsky, M. E. (1991). The Fujita–Ziegler model: a new semi-quantitative technique for estimating paleoclimate from paleogeographic maps. *Palaeogeography, Palaeoclimatology, Palaeoecology*, **86**, 41–66.

Hallam, A. (1984). Continental humid and arid zones during the Jurassic and Cretaceous. *Palaeogeography, Palaeoclimatology, Palaeoecology*, **47**, 195–223.

Hay, W. W., Behensky, J. F., Jr., Barron, E. J., and Sloan, J. L., II (1982). Late Triassic–Liassic paleoclimatology of the proto-central North Atlantic rift system. *Palaeogeography, Palaeoclimatology, Palaeoecology*, **40**, 13–30.

Karpeta, W. R. (1987). The Cretaceous Mbotyi and Mngazana Formations of the Transkei coast: their sedimentological and structural setting. *South African Journal of Science*, **90**, 25–36.

Lloyd, C. R. (1982). The Mid-Cretaceous Earth: paleography; ocean circulation and temperature; atmospheric circulation. *Journal of Geology*, **90**, 393–413.

Mabesoone, J. M., Fúlfaro, V. J., and Suguio, K. (1980). Phanerozoic sedimentary sequences of South American Platform. *Earth Science Reviews*, **17**, 49–68.

McCabe, P. J. and Parrish, J. T. Tectonic and climatic controls on the distribution and quality of Cretaceous coals. In *Controls on the distribution and quality of Cretaceous coals* (ed. P. J. McCabe and J. T. Parrish), Geological Society of America Special Paper (in press).

McKee, E. D. (1979). Introduction to a study of global sand seas. In *A study of global sand seas* (ed. E. D. McKee), *US Geological Survey Professional Paper*, **1052**, 1–19.

McLachlan, I. R. and McMillan, I. K. (1976). Review and stratigraphic significance of southern Cape Mesozoic palaeontology. *Transactions of the Geology Society of South Africa*, **79**, 197–212.

McLachlan, I. R. and Pieterse, E. (1978). Preliminary palynological results: Site 261, Leg 40, Deep Sea Drilling Project. In *Initial Reports of the Deep Sea Drilling Project* (ed. H. M. Bolli *et al.*), Vol. 40, pp. 857–81.

Martin, H. (1973). The Atlantic margin of southern Africa between latitude 17° south and the Cape of Good Hope. In *The ocean basins and margins* (ed. A. E. M. Nairn and F. G. Stehli), Vol. 1. The South Atlantic, pp. 277–300. Plenum Press, New York.

Millot, G. (1970). *Geology of clays*. Chapman and Hall, London.

Morgan, R. (1978). Albian to Senonian palynology of Site 364, Angola Basin. *Initial Reports of the Deep Sea Drilling Project* (ed. H. M. Bolli *et al.*), Vol. 40, 915–51.

Murphy, M. G. and Kennett, J. P. (1985). Development of latitudinal thermal gradients during the Oligocene: Oxygen-isotope evidence from the Southwest Pacific. *Initial Reports of the Deep Sea Drilling Project* (ed. J. P. Kennett *et al.*), Vol. 90, 1347–60.

Nicolas, J. and Bildgen, P. (1979). Relations between the location of the karst bauxites in the northern hemisphere, the global tectonics and the climatic variations during geologic time. *Palaeogeography, Palaeoclimatology, Palaeoecology*, **28**, 205–39.

Nwajide, C. S. (1986). Fabrics of meteoric phreatic diagenesis: inferences from the petrographic analysis of the Turonian Wadatta Limestone, Nigeria. *Journal of African Earth Sciences*, **5**, 641–50.

Parrish, J. M., Parrish, J. T., and Ziegler, A. M. (1986). Permian-Triassic paleogeography and paleoclimatology and implications for therapsid distributions. In *The ecology and biology of mammal-like reptiles* (ed. N. H. Hotton, III, P. D. MacLean, J. J. Roth, and E. C. Roth), pp. 109–32. Smithsonian Press, Washington, DC.

Parrish, J. T. (1982). Upwelling and petroleum source beds, with reference to the Paleozoic. *American Association of Petroleum Geologists Bulletin*, **66**, 750–74.

Parrish, J. T. (1985). Global paleogeography, atmospheric circulation, and rainfall in the Barremian Age (late Early Cretaceous). *US Geological Survey Open-File Report*, **85**, 728.

Parrish, J. T. Jurassic climate and oceanography of the circum-Pacific region. In *The Jurassic of the circum-Pacific* (ed. G. E. G. Westermann). Oxford University Press, Oxford. (In press.)

Parrish, J. T. and Curtis, R. L. (1982). Atmospheric circulation, upwelling, and organic-rich rocks in the Mesozoic and Cenozoic Eras. *Palaeogeography, Palaeoclimatology, Palaeoecology*, **40**, 31-66.

Parrish, J. T. and Doyle, J. A. (1984). Predicted evolution of global climate in Late Jurassic–Cretaceous time. *International Organization of Paleobotany Conference*, Edmonton, Alberta, August, 1984, *Abstracts*.

Parrish, J. T., Gaynor, G. C., and Swift, D. J. P. (1983). Circulation in the Cretaceous Western Interior Seaway of North America—a review. In *Mesozoic of middle North America* (ed. D. F. Stott and D. J. Glass), *Canadian Society of Petroleum Geologists Memoir*, **9**, 211–31.

Parrish, J. T., Ziegler, A. M., and Scotese, C. R. (1982). Rainfall patterns and the distribution of coals and evaporites in the Mesozoic and Cenozoic. *Palaeogeography, Palaeoclimatology, Palaeoecology*, **40**, 67–101.

Patzkowsky, M. E., Smith, L. H., Markwick, P. J., Engberts, C. J., and Gyllenhaal, E. D. (1991), Application of the Fujita–Ziegler paleoclimate model: Early Permian and Late Cretaceous examples. *Palaeogeography, Palaeoclimatology, Palaeoecology*, **86**, 67–85.

Petri, S. and Campanha, V. A. (1981). Brazilian continental Cretaceous. *Earth Science Reviews*, **17**, 69–86.

Prasad, G. (1983). A review of the early Tertiary bauxite event in South America, Africa and India. *Journal of African Earth Sciences*, **1**, 305–13.

Quattrocchio, M. (1982). Sobre el posible significado paleoclimático de los quistes de dinoflagelados en el Jurásico y Cretácico inferior de la Cuenca Neuquina. III Congreso Argentino de Paleontología y Bioestratigrafía. *Actas*, pp. 107–13.

Reyment, R. A. (1981). West Africa. In *Aspects of mid-Cretaceous regional geology* (ed. R. A. Reyment and P. Bengtson), IGCP Project 58, Mid-Cretaceous Events, pp. 133–60. Academic Press, London.

Riccardi, A. C. (1988). *The Cretaceous system of southern South America*. Geological Society of America Memoir, 168.

Robert, C. (1980). Climats et courants cénozoïques dans l'Atlantique Sud d'après l'étude des minéraux argileux (legs 3, 39 et 40 DSDP). *Oceanologica Acta*, **3**, 369–76.

Robinson, P. L. (1973). Palaeoclimatology and continental drift. In *Implications of continental drift to the earth sciences* (ed. D. H. Tarling and S. K. Runcorn), Vol. I, pp. 449–76. Academic Press, London.

Ronov, A. B. and Balukhovskii, A. N. (1982). Climatic zones on continents and the general trend of climatic changes during the late Mesozoic and Cenozoic. *Lithology and Mineral Resources*, **5**, 508–21.

Rowley, D. B., Raymond, A., Parrish, J. T., Lottes, A. L., Scotese, C. R., and Ziegler, A. M. (1985). Carboniferous paleogeographic, phytogeographic, and paleoclimatic reconstruction. *International Journal of Coal Geology*, **5**, 7–42.

Savin, S. M. (1977). The history of the Earth's surface temperature during the past 100 million years. *Annual Review of Earth and Planetary Sciences*, **5**, 319–56.

Shackleton, N. and Boersma, A. (1981). The climate of the Eocene ocean. *Journal of the Geological Society of London*, **138**, 153–8.

Tankard, A. J., Jackson, M. P. A., Eriksson, K. A., Hobday, D. K., Hunter, D. R., and Minter, W. E. L. (1982). *Crustal evolution of Southern Africa. 3.8 Billion years of earth history*. Springer-Verlag, New York.

Van Houten, F. B. (1982). Ancient soils and ancient climates. In *Climate in earth history* (ed. Geophysics Study Committee), pp. 112–17. National Academy of Sciences, Washington, DC.

Van Houten, F. B. and Bhattacharyya, D. P. (1982). Phanerozoic oolitic ironstones–geologic record and facies model. *Annual Review of Earth and Planetary Sciences*, **10**, 441–57.

Volkheimer, W. (1967). Palaeoclimatic evolution in Argentina and relations with other regions of Gondwana. In *Gondwana Stratigraphy*, IUGS Symposium, 1–15 October 1967, Buenos Aires, Vol. 2, pp. 551–87. UNESCO Earth Sciences.

Volkheimer, W. (1971). Aspectos paleoclimatologicos del Terciario Argentino. *Revista del Museo Argentino de Ciencias Naturales 'Bernardina Rivadavia' e*

Instituto Nacional de Investigacion de las Ciencias Naturales, **1**, 244–62.

Volkheimer, W. (1980). Microfloras del Jurásico superior y Cretácico inferior de America Latina. II Congreso Argentino de Paleontología y Bioestratigrafía y I Congreso Latinoamericano de Paleontología, 2–6 April 1978, *Actas*, Vol. II, Buenos Aires, pp. 275–90.

Wise, S.W., Gombos, A. M., and Muza, J. P. (1985). Cenozoic evolution of polar water masses, southwest Atlantic Ocean. In *South Atlantic Paleoceanography* (ed. K. J. Hsü and H. J. Weissert), pp. 283–315. Cambridge University Press, Cambridge.

Wolfe, J. A. (1980). Tertiary climates and floristic relationships at high latitudes in the northern hemisphere. *Palaeogeography, Palaeoclimatology, Palaeoecology*, **30**, 313–23.

Ziegler, A. M., Scotese, C. R., and Barrett, S. F. (1983). Mesozoic and Cenozoic paleogeographic maps. In *Tidal friction and the Earth's rotation II* (ed. P. Brosche and J. Sündermann), pp. 240-52. Springer-Verlag, Berlin.

4 ASPECTS OF AROID GEOGRAPHY

Simon J. Mayo

PRELIMINARIES

Why the aroids?

It would be understandable if the reader were puzzled to find in the present volume a discussion of a family of organisms which shows only weak and distant relationships across the South Atlantic Ocean. There are, after all, many other tropical plant taxa which show striking disjunctions between Africa and South America (Thorne 1973a and Good 1974 give lists of genera). In at least a proportion of these cases, such disjunctions are probably evidence of previous land continuity between the continents. There are, nevertheless, good reasons for considering the geographical patterns shown by taxa which do not fit neatly into the simple vicariance model represented by the tectonic splitting of South America and Africa.

In the first place, the history of the floras since the rifting of South America–Africa took place has probably been much influenced by other factors, such as differing climatic histories. The overall similarity of the floras of the two continents is surprisingly low given such a clear geophysical background (Thorne 1973a). Thorne (1973a) and Smith (1973) considered that the opening of the South Atlantic took place too early in angiosperm evolution (Jurassic–Cretaceous) to have affected the distributions of modern taxa, while Raven and Axelrod (1974) favoured the concept of a uniform pantropical flora in the early Tertiary followed by widespread extinction in Africa as the continent became more arid.

Second, Nelson and Platnick (1980), representing the vicariance biogeography view, argued that biogeographical relationships between two areas can only be evaluated in the context of a minimum of three areas of endemism—the relationship of two areas is meaningful only by reference to a third. This approach requires the investigation of transatlantic biogeography through the study of the phylogeny of widespread continental groups with endemic taxa in each of at least three continental areas. In short, the evaluation of tropical transatlantic relationships in a group of organisms is unlikely to progress far without including consideration of Asia, Australasia, and the Pacific as well.

With these complications in mind, the Araceae are thus seen to be a much more interesting subject for discussion. The family forms a single, easily recognized taxon which has always been regarded as taxonomically isolated, and thus by inference monophyletic (for example Cronquist 1968; Takhtajan 1969; Dahlgren and Clifford 1982; Dahlgren and Rasmussen 1983; and Dahlgren *et al.* 1985). It is a large, subcosmopolitan family, with 2500–3000 species and 105 genera, most of which are endemic to the three major tropical regions of America, Africa, and southeast Asia. These regions can be further subdivided into continental plates—North America, South America, Africa, Madagascar, India, Asia, Papuasia-Australia, etc.–each with endemic taxa of Araceae. On a superficial view, such a family might be expected to show clear signs in its biogeography of the South America–Africa connection.

In the account which follows I have not attempted a strict vicariance analysis, though this would be a desirable future goal. Instead I have tried to reappraise, along fairly traditional lines, the major disjunctions in aroid geography and their taxonomic basis. My aim has been to compare tropical transatlantic distributions with those across the Indian and Pacific Oceans. What emerges is a pattern of relationships that clearly link Africa and South America individually to Asia, but only rather tenuously to each other. While this conclusion cannot be applied generally to the angiosperm floras of Africa and South America it nevertheless represents a pattern which is not uncommon in other plant families.

What are aroids?

Aroids, or Araceae, are herbaceous monocotyledons characterized by their inflorescences, which consist of a fleshy spadix (spike) of minute flowers surrounded by a bract called a spathe (Figs 4.1 and 4.2). Bown (1988) gives an excellent and readable general account of the family. In temperate countries, aroids are familiar both in the wild or as garden plants (for example Lords and Ladies—*Arum maculatum*; Arum Lily—*Zantedeschia aethiopica*) and as ornamental house plants, such as the Swiss Cheese Plant (*Monstera deliciosa*), Flamingo Flower (*Anthurium andraeanum*), dieffenbachia or Leopard Lily (cultivars of *Dieffenbachia seguine*) and philodendrons (for example *Philodendron erubescens*, Fig. 4.2). Aroids are most abundant and diverse in the tropics, where some are important food plants, particularly Taro (*Colocasia esculenta*), Tannia or Cocoyam (*Xanthosoma sagittifolium*) and Elephant Yam (*Amorphophallus paeoniifolius*, Fig. 4.1).

Unlike many monocots, the leaves of aroids are generally broad and pinnately veined. The most common leaf shape is cordiform (heart-shaped)

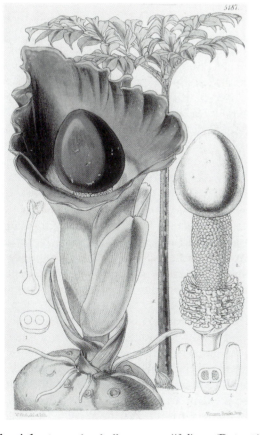

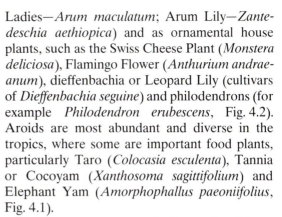

Fig. 4.1 *Amorphophallus paeoniifolius* (Dennst.) Nicolson; a form with a relatively small and smooth sterile spadix appendix; from *Bot. Mag.*, **86**, t.5187 (1860).

Fig. 4.2 *Philodendron erubescens* C. Koch and Augustin; from *Bot. Mag.*, **84**, t.5071 (1858).

or sagittate (arrow-shaped, Fig. 4.2), but greatly variety occurs even within a single genus (for example *Anthurium*, *Philodendron*), and some forms are almost exclusive to the family, for example the leaf of *Amorphophallus* (Fig. 4.1). There is also much diversity in habit and ecology. In temperate lands we are accustomed to think of aroids as geophytes having tubers or rhizomes (for example *Arum*), but in the tropics many, if not most, are forest hemi-epiphytes or lianas (for example *Monstera*, *Philodendron*, *Rhaphidophora*). Aquatic or subaquatic forms are also frequent (for example *Pistia*, *Cryptocoryne*).

The inflorescences of aroids show wide variation within the overall organization of spathe and spadix. Essentially the whole structure behaves like one blossom and the diversity of construction found within the family reflects different levels of specialization of floral organs. The simpler plants (for example *Anthurium*) have unspecialized spathes and uniform spadices composed entirely of bisexual flowers, each surrounded by a whorl of tepals. *Monstera* exemplifes a slightly more specialized type in which the tepals are absent and the spathe is deciduous immediately after pollen is released. In the more complex forms such as *Amorphophallus* (Fig. 4.1) or *Philodendron* (Fig. 4.2), the fertile flowers are distributed in unisexual zones and other parts of the spadix may bear sterile organs of various types. The spathe is also more distinctly differentiated, forming a spreading limb above and an overlapping tube below.

Evolution in aroids

Ideas on aroid evolution have generally followed Engler's scheme (Engler 1920*b*, p. 63) in which the primitive types are assumed to be those most closely resembling the 'typical' monocot flower, i.e. the *Lilium*-like plan, in which the bisexual flower has two whorls of three similar tepals, the stamens are equal in number and opposite the tepals, and the pistil consists of three fused carpels forming a trilocular ovary with several axile ovules in each locule. Examples of aroid genera

which approach this condition are *Pothos* and *Spathiphyllum*. Simple spathes, not greatly differentiated from leaves or reduced leaves (cataphylls) as in *Anthurium*, are assumed to be primitive.

In arranging the genera into what he considered to be, broadly speaking, a phylogenetic classification, Engler essentially postulated two main evolutionary trends from a heterogeneous primitive subfamily (Pothoideae):

1. increasing elaboration of specialized cell types, especially laticifers and trichosclereids;
2. differentiation of distinct zones and organs within the inflorescence, forming a pseudanth (inflorescence acting as a single blossom).

The evolution of tissues with abundant trichosclereids (slender T- or H-shaped cells with thick, lignified walls—see Nicolson 1960) has occurred in only one group, subfamily Monsteroideae. Trichosclereids are large enough to be visible to the naked eye and are highly irritant when handled or ingested due to their needle-like shape. They are probably an adaptation to protect tissues from herbivores. Laticifer systems (latex-containing cells arranged in tube-like series) occur in the tissues of most aroids except the first two of Engler's subfamilies (Pothoideae and Monsteroideae). No specific ecological role has been assigned to them, but it is reasonable to suppose that they provide chemical protection. This inference is supported circumstantially by numerous reports of aroids with antiseptic and poisonous properties (for example Plowman 1969).

The evolution of the inflorescence is regarded as a trend of increasing specialization and differentiation. In primitive aroids like *Pothos* or *Anthurium*, the inflorescence is a simple spike of similar flowers. Advanced types have evolved into a highly organized pseudanth with specialized zones for attracting pollinators by odour and colour, pollen reception and presentation, and protection of pollen and ovules. This has entailed the loss of tepals, sexual differentiation of fertile flowers, specialization of the spathe into tube and limb, and the modification of both spadix and spathe parts into odour- and resin-secreting tissues, food bodies, slip zones, etc. (for example Knoll 1926; Pohl 1931; Van der Pijl 1937; Vogel 1963; Madison 1979).

Engler (1884, 1920*b*), visualized these evolutionary trends of the inflorescence as having occurred in parallel in various phyletic lines and to different degrees. Within the primitive Pothoideae evolution of unisexual naked flowers had taken place (*Culcasia*), whereas in the subfamily Monsteroideae unisexuality had not evolved but a lesser degree of specialization had been reached in the loss of tepals in the advanced tribe Monstereae. Subfamily Lasioideae showed the whole gamut from bisexual tepalate flowers to highly specialized types (*Amorphophallus*). Subfamily Aroideae included the most highly specialized groups (tribe Areae, tribe Cryptocoryneae, tribe Ambrosineae), although its most primitive tribe, Stylochaetoneae, revealed its earlier stage of phyletic development through the possession of a perigon and stamens with filiform filaments.

Grayum (1984), in a major survey of the taxonomic literature, combined with a palynological study of the family, presented a cladogram which differs from Engler's 'tree' (Engler 1920*b*) in major respects. There are, nevertheless, fundamental similarities, since Grayum also accepts the concept of parallel evolutionary trends in the flowers from bisexuality to unisexuality and from presence to absence of a perigon.

Taxonomy of aroids

Engler's classification, completed with the help of K. Krause for *Das Pflanzenreich* (Engler 1920*b* gives a synopsis), remains the basis for aroid taxonomy in the sense of a detailed and comprehensive taxonomic treatment. The arrangement of genera and the generic composition of the tribes was updated by Bogner (1979), providing modern workers with a more practical taxonomic framework. Nicolson (1982) published an English translation of the synoptical keys to the tribes and genera in *Das Pflanzenreich*, and he and

Bogner (Bogner and Nicolson 1991) later modified this considerably to take account of more recent studies. Grayum (1984) proposed a radically different classification, and later (Grayum 1987) removed *Acorus* from the Araceae to its own family, Acoraceae, a decision which has been accepted by most aroid taxonomists.

By and large, however, the limits of the taxa of most concern in this chapter (genera, subtribes, and tribes) have not changed much since Bogner's (1979) update of the Englerian classification; the major problems of aroid taxonomy are at the subfamily level. In the following discussions Bogner (1979) is used as the taxonomic basis unless otherwise indicated.

BIOGEOGRAPHY—CONTINENTAL DISJUNCTIONS

Aroids show disjunctions and vicariance of genera and tribes across all the major ocean basins of tropical latitudes. In this book, those across the Atlantic Ocean are of most immediate interest. However, my aim is to show that, at least in the aroids, transatlantic connections must be seen in the light of patterns which involve the Indian and Pacific Oceans. The latter will, therefore, be reviewed first. Further information and a discussion of aroid biogeography can be found in Engler (1879, 1909, 1920*b*), Croat (1979), and Grayum (1984).

Indian Ocean disjunctions

These may be divided into two types which I call lowland and highland tracks (Croizat *et al.* 1974). The first concerns tropical plants mostly occurring at lower altitudes and is characterized by strong taxonomic divergence between the African and Asian components, suggesting ancient disruption of formerly continuous ranges. The second is shown by three genera found in highland areas in the tropics and at lower altitudes in the subtropics. There is only weak taxonomic divergence between the African and Asian components, suggesting a more recent vicariance event or even long distance dispersal (in *Remusatia*).

Lowland track

This pattern is shown by three genera and one tribe (Table 4.1). All these taxa occur throughout south-east Asia, from India to Papuasia, but their representation in Africa differs in extent.

Of these four taxa, only *Amorphophallus* spreads right across tropical Africa (Fig. 4.3). This genus has its main centre of diversity in south-east Asia, but the African and Madagascan species represent a significant endemic element, comprising several sections and showing distinctive cytological characters (Chauhan and Brandham 1985). In Asia, *Amorphophallus* species diversity is greatest in West Malaysia and Thailand, becoming progressively attenuated towards the east, with just one or two species in New Guinea and Australia.

Rhaphidophora has a more restricted range and fewer species in Africa (Fig. 4.3). It is absent from Madagascar and has a strongly bicentric distribution in the Malaysian archipelago. Papuasia is the major centre of diversity, with about 30 species, but there are numerous endemic species

Table 4.1 Genera and tribes of lowland tropical Araceae disjunct across the Indian Ocean

Taxon	Approximate numbers of species in each region			
	Africa	Madagascar	India, SE Asia, Sunda shelf[1]	Sahul shelf[1], New Guinea, N. Australia
Amorphophallus	26	1	64	3
Rhaphidophora	2	0	72	44*
Pothos	0	1	62	20
Lasieae (tribe)	1	0	7	12*

[1] Ref. Whitmore (1981).
* Includes species from the Solomon Islands, Fiji, Samoa, and other islands of the western Pacific.

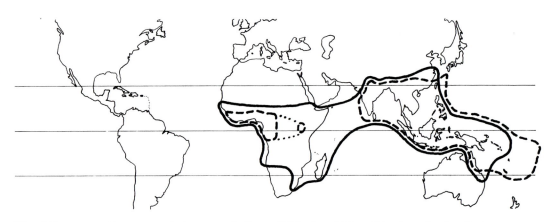

Fig. 4.3 Geographical range of *Amorphophallus* (solid line) and *Rhaphidophora* (broken line).

in continental south-east Asia and the Sunda shelf region as well. Engler and Krause (1908) regarded the two African species as sufficiently distinct from the rest of *Rhaphidophora* to warrant recognition as a separate genus, *Afrorhaphidophora*. In uniting these genera (Bogner 1979), modern taxonomists disagree with this view, but the African component none the less represents an endemic element, characterized morphologically by the presence of basal ovules in the ovary.

Tribe Lasieae (Fig. 4.4) has a similar track to that of *Rhaphidophora* and tribe Monstereae (Fig. 4.4), to which *Rhaphidophora* belongs. Engler (1911) regarded the single African species of Lasieae as a species of *Cyrtosperma*, but recent studies by Hay (1988) support Schott's earlier view (Schott 1860) of a separate genus *Lasimorpha*. As in *Rhaphidophora*, the African component is restricted to the west and centre of the continent and is a distinctive endemic taxon.

Pothos is entirely Asian except for its occurrence in Madagascar (Bogner 1975). The Madagascan element is *P. scandens*, a common species in India, Ceylon (Nicolson 1988), and south-east Asia. This disjunction differs from the others described above by absence from continental Africa and the lack of taxonomic divergence between the two parts of the range, suggesting the possibility of long-distance dispersal. Nevertheless, the Madagascar–Asia track occurs in many other genera (Good 1974) and the possibility that *P. scandens* is an ancient component of Madagascar's flora therefore cannot be ruled out.

Highland track

The three aroid genera showing this distribution are *Arisaema*, *Sauromatum*, and *Remusatia*. Although not closely related taxonomically, they are similar in having their main centre of diversity in the area bounded by Nepal in the west and the Chinese province of Yunnan in the east. Taxonomic divergence between African and Asian components is weak in all three genera. The track is made up either of species disjunctions or vicariant species pairs. Cases of the former are *Sauromatum venosum* (Fig. 4.5; see also Mayo 1985), *Arisaema flavum*, and *Remusatia vivipara*. The latter species may be a case in which long-distance dispersal has played an important part in distribution due to a curious mode of vegetative propagation that *Remusatia* shares with the related Himalayan genus *Gonatanthus* (Engler and Krause 1920). The tuber produces erect or stoloniferous axes with clusters of small bubils, each bearing recurved scale leaves. The bulbils detach readily and it is easy to visualize their transport by birds, once attached to their plumage.

Vicariant species pairs occur in *Arisaema* between highland areas of Ethiopia, Yemen, and northern and southern India (Mayo and Gilbert 1986). This genus shows a similar type of

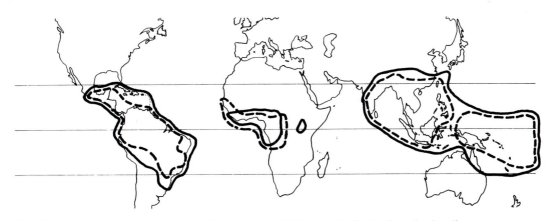

Fig. 4.4 Geographical ranges of tribe Monstereae (solid line) and tribe Lasieae (broken line).

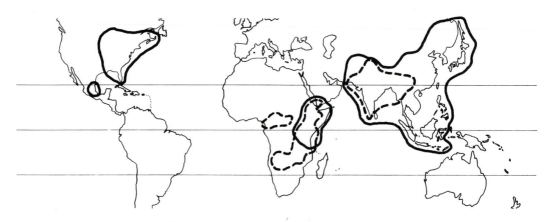

Fig. 4.5 Geographical ranges of *Arisaema* (solid line) and *Sauromatum* (broken line).

vicariance between north-east Asia and eastern North America and in its overall distribution (Fig. 4.5) has much in common with the more widespread genus *Impatiens* (Balsaminaceae; see Grey-Wilson 1980).

Transpacific disjunctions

The transpacific connections of Araceae are the most striking biogeographical feature of the family and have been discussed by many authors (for example Engler 1920*b*; Bunting 1960; Van Steenis 1962, 1979; Whitmore 1969; Nicolson 1968). Two main types may be distinguished, one in the tropics and the other in the north temperate

zone. As in the case of the Indian Ocean disjuncts, the tropical track is characterized by stronger taxonomic divergence.

Tropical track

Of the six taxa exhibiting this pattern (Table 4.2), the two tribes Lasieae (Fig. 4.4) and Monstereae (Fig. 4.4) are pantropical and the remainder are absent from Africa. Subfamily Colocasioideae is represented in Africa only by *Remusatia vivipara*, discussed earlier. *Homalomena* and *Schismatoglottis* have rather similar distributions (Fig. 4.6). Both genera are bicentric in Malesia with many endemic species in both Papuasia and west Malaysia. In tropical America they are each

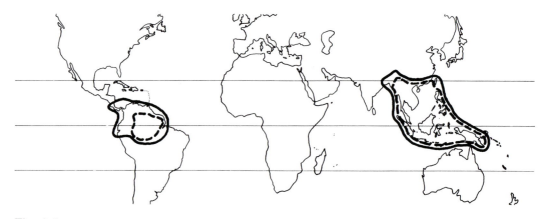

Fig. 4.6 Geographical ranges of *Homalomena* (solid line) and *Schismatoglottis* (broken line).

Table 4.2 Taxa of tropical Araceae disjunct across the Pacific Ocean

Taxon	Approximate numbers of species in each region		
	Tropical America	India, SE Asia, Sunda shelf[1]	Sahul shelf[1], New Guinea, W. Pacific[2]
Homalomena	9*	77	28
Schismatoglottis	3	102	20
Spathiphyllum	38	1	3
Lasieae (tribe)	39	7	12
Monstereae (tribe)	84	115	66
Colocasioideae[3] (subfamily)	111	80	16

* M. Moffler, personal communication
[1] Ref. Whitmore (1981)
[2] Includes N. Australia, Solomon Islands, Fiji, Samoa, and other islands
[3] *Protarum* not included

represented by a distinct section consisting of only a few species.

In the Monstereae and Lasieae tribes (Fig. 4.4) there are no disjunct genera, but in the Monstereae, the two pairs of genera *Monstera—Epipremnum* and *Alloschemone–Scindapsus* may be cited as possible examples of vicariance (Madison 1976, 1977).

Spathiphyllum has an especially interesting distribution, previously discussed by Bunting (1960) and Nicolson (1968) among others (Fig. 4.7). Of the four sections recognized by Bunting (1960), only sect. *Massowia* is involved in the transpacific disjunction. The other three are endemic to tropical America. On the Asian side of the Pacific Ocean, sect. *Massowia* has not been found further west than the Celebes. There is an endemic species in the Solomon Islands and a closely related genus, *Holochalmys*, in New Guinea and New Britain. The Pacific disjunction itself

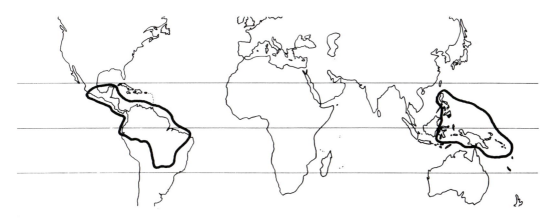

Fig. 4.7 Geographical range of *Spathiphyllum*.

involves two species, *S. commutatum* and *S. cannifolium*, which are so closely similar as to be very difficult to distinguish without knowing the origin of the plants examined. The Papuasia–America link shown by *Spathiphyllum* is mirrored by the generic pair *Epipremnum-Monstera*; *Epipremnum* has its centre of maximum diversity in Papuasia.

Transpacific aroid disjunctions thus link both continental Asia and the Australasian plate to tropical America. Von Steenis (1962) has previously pointed out the predominance of rainforest plants among the transpacific taxa exhaustively reviewed by him, and the same is true of the aroids mentioned here. It is thus difficult to imagine how such distributions could have come about without invoking either rafting of continental plates or humid tropical land continuity in the past.

North temperate track

These disjunctions (Table 4.3) all involve closely related or identical species. *Arisaema* (Fig. 4.5) and *Symplocarpus* form part of the well-known pattern connecting east temperate Asia and eastern North America. *Lysichiton* has a typical trans-Beringia distribution. The taxonomy suggests fairly recent disruption of continuous boreal ranges.

Table 4.3 Taxa of temperate Araceae disjunct across the northern Pacific Ocean

	Approximate numbers of species in each region		
	East Asia	North America	
		West	East
Arisaema	59	0	3
Calla	1	1*	1
Lysichiton	1	1	0
Symplocarpus	1?2	0	1

* Not found west of the Rocky Mountains

Transatlantic disjunctions

The distribution of tribes Lasieae and Monstereae have already been discussed in connection with Indian and Pacific Ocean tracks. Although these tribes, being pantropical, are disjunct across the Atlantic, their genera provide no examples of vicariant pairs separated by this ocean.

Madagascar shows two aroid links to the New World. Bogner (1972) has previously discussed the possible link between tribes Arophyteae (Madagascar) and Spathicarpeae (South America). He demonstrated the closer affinity of the Arophyteae to the Spathicarpeae (Fig. 4.8) than to the neighbouring tropical African tribe Stylochaetoneae. A strong case can be made for linking Madagascan *Typhonodorum* to North American *Peltandra* rather than to southern African *Zantedeschia* (Fig. 4.9, Table 4.4; see also discussion in Grayum 1984).

Further links may be suggested here. There is a striking resemblance between the tropical African genus *Callopsis* and the Amazonian genera of tribe Zomicarpeae (Fig. 4.9). Brown (1901) drew attention to the similarity of *Callopsis* and *Zomicarpella*, although Engler (1920a) later placed them in separate tribes. The genus *Ulearum* (also of tribe Zomicarpeae) seems even closer to *Callopsis*, morphologically.

French (personal communication) has suggested a relationship between neotropical *Philodendron* and the endemic African genera *Culcasia* and *Cercestis* (the latter including *Rhektophyllum*), based on anatomical studies. Following up this idea, Mayo (1986) found similarities in floral morphology and anatomy between *Cercestis mirabilis* (syn. *Rhektophyllum mirabile*) and *Philodendron* subgen. *Meconostigma*.

DISCUSSION

Transatlantic tracks are only weakly expressed in aroids. The disjunctions of tribes Lasieae and Monstereae, and the vicariant genera *Peltandra* and *Typhonodorum* offer the clearest examples.

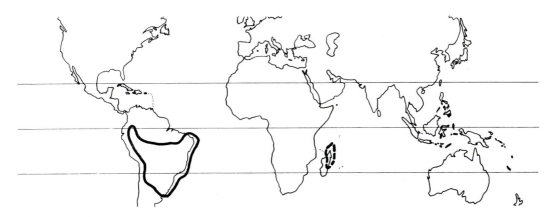

Fig. 4.8 Geographical range of tribe Spathicarpeae (solid line) and tribe Arophyteae (broken line).

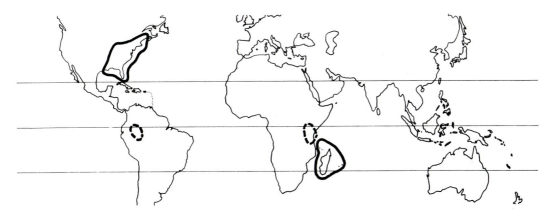

Fig. 4.9 Geographical ranges of *Peltandra* (solid line, North America) and *Typhonodorum* (solid line, Madagascar), and of *Ulearum* (broken line, South America) and *Callopsis* (broken line, East Africa).

However, it seems equally plausible to view these distributions as running east from Africa and Madagascar to Asia, and west from the Americas to Asia. In the case of *Peltandra* and *Typhonodorum* this could mean that closely related but extinct taxa existed in Asia. *Philodendron limnestis*, an Eocene aroid fossil from Tennessee reported by Dilcher and Daghlian (1977) may be relevant to this question. Although ascribed to a tropical American genus by its authors, it seems more likely that it is related to *Typhonodorum* (see discussion in Mayo 1991). Such a conclusion, even if correct, would contribute little to

resolving the problem of the ancient range of the taxonomic alliance represented by these two genera, since *Peltandra* is found today throughout eastern North America. However, it would indicate that related taxa, very similar to *Typhonodorum*, existed in this region in the first half of the Cenozoic.

Cladistic studies of tribes Lasieae and Monstereae are needed to provide a basis for analysing their transoceanic disjunctions. From present evidence, the readily discernible generic links are across the Indian and Pacific Oceans rather than the Atlantic. There is a possibility that *Callopsis*

Table 4.4 Character comparison of three genera of subfamily Philodendroideae

Character	*Peltandra*	*Typhonodorum*	*Zantedeschia*
Spathe	differentiated into tube and limb	differentiated into tube and limb	undifferentiated
Spadix appendix	present or absent	present	absent
Male flower	4–8 connate stamens	4–8 connate stamens	2–3 free stamens
Female flower	staminodes present, connate or free	staminodes present, free	staminodes present or absent, free
Ovary	1-locular	1-locular	2–5-locular
Placentation	basal	basal	axile
Ovules	1–3, amphi- to orthotropous	1–2, orthotropous	1–8 per locule anatropous
Seeds	endosperm sparse	endosperm absent	endosperm present
Leaf proanthocyanidins[1]	present	present	absent
Pollen type[2]	echinulate or reticulate	scabrous	smooth or scabrous

Data from Engler (1915) supplemented by personal observation of material at the Kew Herbarium (K), except for the last two characters.
[1] Data from Williams *et al.* (1981)
[2] Data from Thanikaimoni (1969)

and *Ulearum* represent a tropical Africa–Amazonia link of a type well known in other angiosperm families (for example *Raphia* and *Elaeis* in the palms; see Moore 1973 and Thorne 1973*a* for other examples).

The two types of Indian Ocean track seen in aroids seem to require different phases of continuity between humid forest habitats in Africa and Asia. The lowland track appears to have resulted from an ancient vicariance event which disrupted distributions of taxa extending between Africa and Asia. Since then it seems that evolution in *Amorphophallus*, *Rhaphidophora*, and tribe Lasieae has proceeded independently in the two areas over a period sufficiently long to have produced sectional taxonomic differences. When could such a vicariance event have taken place? According to a survey of fossil seeds of Araceae by Madison and Tiffney (1976), Monstereae and Lasieae formed part of the Tertiary palaeotropical flora of Europe at least as far back as the Oli-

gocene. It seems, therefore, that the Laurasian range of these taxa was then considerably greater than it is now. Continuity with Africa could have occurred with the closure of the Tethys Sea in the Late Cretaceous (Parrish, this volume). The disruption of this continuity could have occurred later as a result of the physical separation of Africa and Eurasia after the Palaeocene (Raven and Axelrod 1974). Connection between the aroid genera would not have been reestablished after the Oligocene collision of the two continents (Parrish, this volume) due to the gradual drying out of northern Africa in the Mid and Late Tertiary (Raven and Axelrod 1974). This explanation reconciles the absence of *Amorphophallus* and *Rhaphidophora* from South America with the apparent need in these taxa for a long history in Africa. By the Late Cretaceous, South America would have been separated from Africa by a wide ocean gap (Parrish, this volume).

The highland trans-Indian Ocean track suggests a more recent phase of vicariance events, possibly during the Pleistocene. Flenley (1979) reviews palynological evidence for vegetation and climate change in tropical Africa during the Quaternary, demonstrating considerable altitudinal movement of the montane forest belt in the last 25 000 years. This suggests that perhaps during earlier interglacial periods a subtropical forest connection between the Himalaya and north-east Africa would have been possible. Friis (1983) has discussed a number of Ethiopian plant distributions including types comparable to the aroid highland track. African aroids thus seem to be rather more strongly connected to Asian taxa than to anywhere else. This certainly corresponds to the situation in many other flowering plant families (for example Thorne 1973a,b; Good 1974 Appendix; Van Steenis 1979; Grey-Wilson 1980; Faden 1983).

The strong aroid links across the Pacific serve to emphasize how important the history of this ocean basin is to understanding the biogeography of pantropical taxa. Van Steenis' (1962) comprehensive review of Pacific plant disjunctions and many important contributions since (for example Van Steenis and Van Balgooy 1966; Whitmore 1969; Van Balgooy 1971; Van Steenis 1979) provide numerous other examples similar to the aroid distributions discussed here. Melville (1981) and Nur and Ben-Avraham (1981) postulated a former Pacific continent to account for these and related biogeographical phenomena. This hypothesis visualizes the rafting of continental plates east to the Americas and west to Asia and Australasia, resulting from the break-up of a putative continent called 'Pacifica'. In another guise, these ideas follow Van Steenis' (1962) concept of a 'land bridge', in requiring some form of land continuity in the Pacific tropics, even if interrupted in time and space, to explain trans-Pacific disjunctions at these latitudes.

For aroids the 'Pacifica' hypothesis is attractive and to some extent supported by the distributions of genera within the neotropics. Most genera are concentrated in the north-western part of South America, Costa Rica, and Panama, becoming rapidly attenuated in species and sectional diversity elsewhere (for example *Monstera*, Fig. 4.10). A few taxa do not fit this pattern and occur to the south and east of South America (for example *Philodendron* subgen. *Meconostigma*, see Mayo 1991; tribe Spathicarpeae, Fig. 4.8; *Zomicarpa*). There are no genera endemic to the West Indies or tropical North America. The aroid taxa involved in the tropical transpacific patterns (Table 4.2) all have the 'Monstera type' range pattern. Madison (1977) and Croat (1982) have previously commented on the respective distributions of *Monstera* and *Syngonium*, noting their species and sectional concentration in Costa Rica and Panama.

The Pacifica hypothesis would suggest that such groups could have been derived from ancestors that arrived in Central America or north-western South America from the west, on

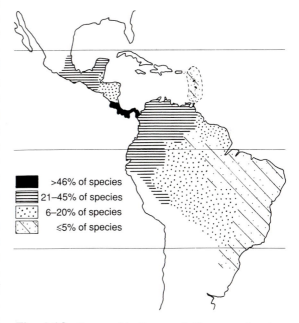

Fig. 4.10 Geographical range of *Monstera*, showing degree of species diversity based on political boundaries; black represents areas containing more than 46 per cent of total species, horizontal lines 21–45 per cent of species, heavy stippling 6–20 per cent of species; light stippling with diagonal hatching represents areas with less than 5 per cent of species.

continental fragments, and later spread further into South America during the Cenozoic. The 'southeastern' aroid taxa of South America could be interpreted as elements of an ancient South American biota that has evolved in isolation in the period between the rifting from Africa and the (putatively later) collision with 'Pacifica' fragments.

Madison (1977) suggests an alternative explanation for the transpacific disjunction represented by *Monstera* and its Asian vicariant partner *Epipremnum*. They may represent part of a Tertiary boreotropical flora that migrated southwards in the Pliocene-Pleistocene with the establishment of land continuity between North and South America. Van der Hammen and Cleef (1983) list other angiosperm genera, and adduce palynological evidence in support of a similar explanation for trans-pacific disjunctions.

To sum up, the relative lack of transatlantic connections in aroids is puzzling. The family is diverse and widespread. It has endemic elements in various regions of the earth which were once parts of Laurasia and Gondwanaland. A long history over a Pangaeic area, stretching back at least to the Cretaceous, thus seems reasonable. Africa, Asia, and the America all have well-defined endemic genera and tribes. Superimposed on this pattern are the transoceanic tracks linking Africa and South America to Asia, but only very tenuously to each other. But if aroids have an ancient Cretaceous history in Africa, why are the links with South America so weak? Perhaps there was a climatic barrier in the Cretaceous to the exchange of ancestral aroids between protoAfrica and proto-South America. Perhaps the geological and tectonic events which gave rise to the other transoceanic disjunctions postdated the rifting of South America and Africa. Perhaps both evolution and extinction of aroids in Africa have proceeded much more rapidly than elsewhere in the tropics due to more intense climatic change (for example Tertiary aridification, an explanation favoured by Raven and Axelrod 1974). There is doubtless no single explanation. In a large and diverse family, it is to be expected that a variety of

geographical patterns will be found that reflect the varying histories of the different phyletic lines within the group. Nevertheless, an overall pattern corresponding to the earth's history should, theoretically at least, become discernible if the group is monophyletic. Such a possibility should be a great stimulus to taxonomic and phylogenetic studies in aroids as well as in other groups of organisms.

ACKNOWLEDGEMENTS

I am grateful to the following colleagues for their comments and advice: J. Bogner, P. Boyce, J. Dransfield, W. George, C. Humphries, G. Lewis, M. Madison, D. Nicolson, R. Polhill, C. Stirton, and T. Whitmore.

REFERENCES

Bogner, J. (1972). Revision der Arophyteae (Araceae). *Botanische Jahrbücher*, **92**(1), 1–63. Stuttgart.

Bogner, J. (1975). *Flore de Madagascar et des Comores: 31e. familie. Aracées*, 75 pp. Museum National d'Histoire Naturelle, Paris.

Bogner, J. (1979). A critical list of the aroid genera. *Aroideana*, **3**(3), 63–73.

Bogner, J. and Nicolson, D. H. (1991). A revised classification of *Araceae* with dichotomous keys. *Willdenowia*, **21**, 35–50.

Bown, D. (1988). *Aroids—plants of the Arum Family*, 255 pp., Century, London.

Brown, N. E. (1901). Aroideae. In *Flora of tropical Africa*, Vol. 8 (ed. W. T. Thiselton-Dyer), pp. 137–200. Lovell Reeve, London.

Bunting, G. S. (1960). A revision of Spathiphyllum. *Memoirs of the New York Botanical Garden*, **10**(3), 1–53.

Chauhan, K. P. S. and Brandham, P. E. (1985). Chromosome and DNA variation in *Amorphophallus* (Araceae). *Kew Bulletin*, **40**(4), 745–58.

Croat, T. B. (1979). The distribution of Araceae. In *Tropical Botany*, (ed. K. Larsen and L. B. Holm-Nielsen), pp. 291–308. Academic Press, London.

Croat, T. B. (1982). A revision of *Syngonium* (Araceae). *Annals of the Missouri Botanical Garden*, **68**(4), 565–651.

Croizat, L., Nelson, G., and Rosen, D. E. (1974). Centers of origin and related concepts. *Systematic Zoology*, **23**(2), 265–87.

Cronquist, A. (1968). *The evolution and classification of flowering plants*, 396 pp. Nelson, London.

Dahlgren, R. T. and Clifford, H. T. (1982). *The Monocotyledons: a comparative study*, 378 pp. Academic Press, London.

Dahlgren, R. M. T. and Rasmussen, F. N. (1983). *Monocotyledon evolution, characters and phylogenetic estimation*. In *Evolutionary biology*, Vol. 16 (ed. M. K. Hecht, B. Wallace, and G. T. Prance), pp. 255–395. Plenum Press, New York.

Dahlgren, R. T., Clifford, H. T., and Yeo, P. F. (1985). *The Families of the monocotyledons, structure, evolution and taxonomy*, 520 pp. Springer-Verlag, Berlin.

Dilcher, D. L. and Daghlian, C. P. (1977). Investigations of angiosperms from the Eocene of southeastern North America: *Philodendron* leaf remains. *American Journal of Botany*, **64**(5), 526–34.

Engler, A. (1879). Araceae. In De Candolle, A. & C., *Monographiae Phanerogamarum*, **2**, 36–55. Masson.

Engler, A. (1884). Beiträge zur Kenntniss der Araceae V. 12. Über den Entwicklungsgang in der Familie der Araceen und über die Blütenmorphologie derselben. *Botanische Jahrbücher*, **5**, 141–88, 287–336.

Engler, A. (1909). Die Bedeutung der Araceen für die pflanzengeographische Gliederung des tropischen und extratropischen Ostasiens. *Sitzungsberichte der Preussische Akademie der Wissenschaften zu Berlin*, **1909**(52), 1258–81.

Engler, A. (1911). Araceae—Lasioideae. In *Das Pflanzenreich*, Vol. 48 (IV.23C) (ed. A. Engler), 130 pp. Engelmann, Leipzig.

Engler, A. (1915). Araceae – Philodendroideae – Anubiadeae – Aglaonemateae, – Dieffenbachieae, – Zantedeschieae, – Typhonodoreae, – Peltandreae. In *Das Pflanzenreich*, Vol. 64 (IV.23Dc) (ed. A. Engler), 78 pp. Engelmann, Leipzig.

Engler, A. (1920*a*). Araceae—Aroideae—Pistioideae. In *Das Pflanzenreich*, Vol. 73 (IV.23F) (ed. A. Engler), 274 pp. Engelmann, Leipzig.

Engler, A. (1920*b*). Araceae—Pars generalis et index familiae generalis. In *Das Pflanzenreich*, Vol. 74 (IV.23A) (ed. A. Engler), 71 pp. Engelmann, Leipzig.

Engler, A. and Krause, K. (1908). Araceae—Monsteroideae. In *Das Pflanzenreich*, Vol. 37 (IV.23B) (ed. A. Engler), pp. 4–149. Engelmann, Leipzig.

Engler, A. and Krause, K. (1920). Araceae—Colocasioideae. In *Das Pflanzenreich*, Vol. 71 (IV.23E) (ed. A. Engler), pp. 3–132. Engelmann, Leipzig.

Faden, R. B. (1983). Phytogeography of African Commelinaceae, *Bothalia*, **14**(3, 4), 553–7.

Flenley, J. R. (1979). *The Equatorial Rain Forest: a geological history*, 162 pp. Butterworths, London.

Friis, I. (1983). Phytogeography of the tropical northeast African mountains. *Bothalia*, **14**(3, 4), 525–32.

Good, R. (1974). *The geography of the flowering plants* (4th edn). Longman, London.

Grayum, M. H. (1984). Palynology and Phylogeny of the Araceae. Ph.D. diss., 852 pp., University of Massachusetts, Amherst.

Grayum, M. H. (1987). A summary of the evidence and arguments supporting the removal of *Acorus* from the Araceae. *Taxon*, **36**(4), 723–9.

Grey-Wilson, C. (1980). *Impatiens of Africa*, 235 pp. Balkema, Rotterdam.

Hay, A. (1988). *Cyrtosperma* (Araceae) and its Old World allies. *Blumea*, **33**, 427–69.

Knoll, F. (1926). Die Arum-Blütenstände und ihrer Besucher (Insekten und Blumen IV). *Abhhandlungen der zoologish-botanischen Gesellschaft Wien*, **12**, 379–481.

Madison, M. (1976). *Alloschemone* and *Scindapsus* (Araceae). *Selbyana*, **1**(4), 325–7.

Madison, M. (1977). A Revision of *Monstera* (Araceae). *Contributions from the Gray Herbarium of Harvard University*, **207**, 3–100.

Madison, M. (1979). Protection of developing seeds in neotropical Araceae. *Aroideana*, **2**(2), 52–61.

Madison, M. and Tiffney, B. H. (1976). The seeds of the Monstereae: their morphology and fossil record. *Journal of the Arnold Arboretum*, **57**(2), 185–204.

Mayo, S. J. (1985). Araceae. In *Flora of tropical East Africa* (ed. R. M. Polhill), 71 pp. Balkema, Rotterdam.

Mayo, S. J. (1986). Systematics of *Philodendron* Schott (Araceae) with special reference to inflorescence characters. Ph.D. thesis, 972 pp., University of Reading.

Mayo, S. J. (1991). A revision of *Philodendron* subgenus *Meconostigma (Araceae)*. *Kew Bulletin*, **46**(4), 601–81.

Mayo, S. J. and Gilbert, M. (1986). A preliminary revision of *Arisaema* (Araceae) in tropical Africa and Arabia. *Kew Bulletin*, **4**(2), 261–78.

Melville, R. (1981). Vicarious plant distributions and paleogeography of the Pacific region. In *Vicariance biogeography: a critique* (ed. G. Nelson and D. E. Rosen), pp. 238–74. Columbia University Press, New York.

Moore, H. E., Jr. (1973). Palms in the tropical forest ecosystems of Africa and South America. In *Tropical forest ecosystems in Africa and South America: a comparative review* (ed. B. J. Meggers, E. S. Ayensu, and W. D. Duckworth), pp. 63–88. Smithsonian Institute Press, Washington, DC.

Nelson, G. and Platnick, N. I. (1980). A vicariance approach to historical biogeography. *Bioscience*, **30**, 339–43.

Nicolson, D. H. (1960). The occurrence of trichosclereids in the Monsteroideae (Araceae). *American Journal of Botany*, **47**, 598–602.

Nicolson, D. H. (1968). The genus *Spathiphyllum* in the East Malesian and West Pacific islands (Araceae). *Blumea*, **16**(1), 119–21.

Nicolson, D. H. (1982). Translation of Engler's classification of Araceae with updating. *Aroideana*, **5**(3), 67–88.

Nicolson, D. H. (1988). Araceae. In *A Revised Handbook to the Flora of Ceylon*, Vol. 6 (ed. M. D. Dassanayake and F. R. Fosberg), pp. 17–101. Balkema, Rotterdam.

Nur, A. and Ben-Avraham, Z. (1981). Lost Pacifica continent: a mobilistic speculation. In *Vicariance biogeography: a critique* (ed. G. Nelson and D. E. Rosen), pp. 341–58. Columbia University Press, New York.

Parrish, J. T. (this volume). The palaeogeography of the opening South Atlantic.

Plowman, T. (1969). Folk uses of New World aroids. *Economic Botany*, **23**(2), 97–122. New York.

Pohl, F. (1931). Anatomische und ökologische Untersuchungen am Blütenstande von *Philodendron selloum* Schott, mit besonderer Berücksichtigung der Harzkanäle und der Beschaffenheit der Pollenkittstoffe. *Planta*, **15**, 506–29.

Raven, P. H. and Axelrod, D. I. (1974). Angiosperm biogeography and past continental movements. *Annals of the Missouri Botanical Garden*, **61**(3), 529–673.

Schott, H. W. (1860). *Prodromus systematis aroidearum*, 602 pp. Typis congregationis machitharisticae, Vienna.

Smith, A. C. (1973). Angiosperm evolution and the relationship of the floras of Africa and America. In *Tropical forest ecosystems in Africa and South*

America: a comparative review (ed. B. J. Meggers, E. S. Ayensu, and W. D. Duckworth), pp. 49–61. Smithsonian Institution Press, Washington.

Takhtajan, A. (1969). *Flowering plants. Origin and dispersal* (trans. C. Jeffrey), 310 pp. Oliver & Boyd, Edinburgh.

Thanikaimoni, G. (1969). Esquisse palynologique des Aracées. *Travaux de la section scientifique et technique. Institut Français de Pondichéry*, **5**(5), 31 pp.

Thorne, R. F. (1973a). Floristic relationships between tropical Africa and tropical America. In *Tropical forest ecosystems in Africa and South America: a comparative review* (ed. B. J. Meggers, E. S. Ayensu, and W. D. Duckworth), pp. 27–61. Smithsonian Institution Press, Washington.

Thorne, R. F. (1973b). Major disjunctions in the geographical ranges of seed plants. *Quarterly Review of Biology*, **47**, 365–411.

Van Balgooy, M. M. J. (1971). Plant geography of the Pacific. *Blumea Suppl.*, **6**, 1–222.

Van der Hammen, T. and Cleef, A. (1983). *Trigonobalanus* and the tropical amphi-pacific element in the North Andean forest. *Journal of Biogeography*, **10**, 437–40.

Van der Pijl, L. (1937). Biological and physiological observations on the inflorescence of *Amorphophallus*.. *Recueil des Travaux botaniques néerlandais*, **34**, 157–67. Nimègue.

Van Steenis, C. G. G. J. (1962). The land-bridge theory in botany. *Blumea*, **11**, 235–542.

Van Steenis, C. G. G. J. (1979). Plant geography of East Malesia. *Botanical Journal of the Linnean Society*, **79**, 97–178.

Vogel, S. (1963). Kapitel 3. Die Duftdrüsen der Araceen. In *Duftdrüsen im Dienste der Bestäubung: über Bau und Funktion der Osmophoren* (ed. S. Vogel), pp. 639–77. Abhandlungen der Mathematisch-naturwissenchatlichen Klasses Akademie der Wissenchaften und der Litteratur, Mainz.

Whitmore, T. C. (1969). Geography of the flowering plants (of the Solomon Islands). *Philosophical Transactions of the Royal Society*, B255, 549–66.

Whitmore, T. C. (1981). Wallace's Line and some other plants. In *Wallace's Line and plate tectonics* (ed. T. C. Whitmore), pp. 70–80. Clarendon Press, Oxford.

Williams, C. A., Harborne, J. B., and Mayo, S. J. (1981). Anthocyanin pigments and leaf flavonoids in the family Araceae. *Phytochemistry*, **20**(2), 217–34.

5 AFRICAN–AMERICAN RELATIONSHIPS IN THE ACRIDIANS (INSECTA, ORTHOPTERA)

C. Amedegnato

The first fossils resembling insects, the Collembols, date back to the Devonian. The first fossils of well diversified true insects are from the Carboniferous. In this chapter we are mainly concerned with the Order Orthoptera, an ancient order which shows many primitive characters. It occurs as early as the Carboniferous, bearing the same morphological features as now. It is apparent from known fossils that the diversity of phyla, most of which survive to the present day, has been very great, the major extant groups having originated at the beginning of the Palaeozoic.

Today, the most abundant group of modern Orthoptera is represented by the acridians, in their general sense (essentially comprising the Caelifera but excluding the Tetrigoidea and the Tridactyloidea).[1] A phytophagous way of life, widely adapted to all environments, has made this group one of the dominant entomofauna's in terms of biomass, both on land and in some aquatic habitats. Characteristic fossils, whether of recent families or of an ancestral group (Locustopseidae) are found from the Triassic onwards. Their diversification seems to go back to the Jurassic and may be contemporary with the appearance of the angiosperms. This antiquity should therefore be directly represented in the distribution of the phyla relative to the great continental plates and their movements.

In reality, the present day distribution of the group is a direct result both of its antiquity and of its ecological characteristics. Although mainly tropical, acridians are found in many different environments where they display a great variety of ecotypes and extremely varied dispersal abilities,

ranging from inexistent to the abilities illustrated by the great migratory acridians (locusts). The migratory acridians have drawn the attention of so many authors that this phylum has often been unconsciously reduced to a single ecotype which has thus dominated the biogeographical view of the group. Nevertheless, for specialists who study a particular family and its evolution, the characters of antiquity, of ecological versatility, and of dispersal ability are those that have provided a basis for analysis, particularly for the interpretation of the faunal relationships between continents. Although the problem of relationships between Africa and America have always been of interest, sound knowledge is lacking because, in the absence of adequate evidence, the evolution of the group has remained poorly understood. However, in the last few years several studies, even if in need of some reassessment, have improved the situation and we should now begin to understand the subject better.

It is also now known that most of the acridians, in a general sense, show weak to moderate dispersal abilities, and as a matter of fact are distributed relatively passively, following variations in the vegetation, as is natural for phytophages. Obviously this also implies that for their dispersal, acridians usually require land bridges, or, at the most, minor barriers for their dispersal.

Age, evolution, ecological potential, and dispersal ability: all these are key-points which must be considered in order to make the best use of the known distribution patterns and to shed light on the problem of continental relationships.

Ecological and dispersal potentials are the means by which phyla spread and will be considered first. Age and evolution must be examined

[1] A grouping which is conveniently and commonly called Acridomorpha (Dirsh 1966).

simultaneously with phylum distribution, although we must be extremely cautious in any interpretation. This will enable us to draw together our knowledge of the relationship between Africa and America as shown by acridians.

DISPERSAL ABILITIES OF THE ACRIDOMORPHA

Two main factors control dispersal: the potential for communication (the presence of barriers or of land routes), modified by a second factor, the ecology of the species, which may substantially alter the significance or efficacy of these barriers and communication routes.

The ecology of acridians

Acridian ecology is dominated by a close adaptation to the environment, and by the production of ecological forms adapted to living on the ground, to various kinds of herbaceous or arboreal vegetation, or to stable or unstable environments (migratory forms). Broadly speaking it is a classic ecological case of primary consumers which are strictly dependent on the immediate environment. This is evidenced by the close matching of the species distribution to that of the vegetation (see Otte 1981, 1984; Descamps 1984). However, at higher taxonomic levels, which can extend from subfamily to family, there is a highly varied ecological range covering a wide spectrum of biotopes of the continent concerned. In part, this has ensured the survival of the phyla through the climatic changes which have occurred in the course of geological time.

Depending on the circumstances, acridians thus vary from little to highly mobile, from mono- to polyphage, and this determines the dispersal ability of the species.

Role of barriers in the biogeography of acridians

Mountains, rivers, and seas do not constitute equivalent barriers in the dispersal of wingless bush-living species, of winged geophiles, or of prolific winged species, adapted for long migrations or, on the other hand, in the dispersal of insects with limited reproductive potential restricted to a small microbiotope in the canopy of tropical trees.

However, even for the large winged forms, the great mountain ranges of the world constitute a near-complete barrier to dispersal. The northern Andes separate the fauna of Central and South America. The Mexican Altiplano is almost exclusively populated by an ancient fauna and a nearctic fauna, while the lowlands on each side are inhabited by a high proportion of forms which are neotropical in origin. The Himalayas constitute an efficient barrier to the expansion of even the largest migratory forms. In the same way, the East African Rift Valley is a critical part of the biogeography of that continent (see Descamps 1977).

Although fragile, unstable, and narrow, rivers nevertheless constitute, albeit on a smaller scale, very substantial obstacles which may take a long time to be crossed. The shape of the foci of forest dispersal in South America is an example. These dispersal centres, with their characteristic faunas, resulted from the expansion of an original Pleistocene assemblage up to an impassable physical barrier. At present, the Amazonian dispersal centres are limited by great rivers: the Amazon, the Rio Negro, and the Rio Madeira (Müller 1973; Simpson and Haffer 1978; Amedegnato and Descamps 1982).

Hence, if river barriers can be effective, then marine ones must be even more so. Indeed, apart from a few exceptional cases, almost no dispersal of acridians has occurred across marine barriers (at least not in sufficient numbers to constitute a colonization event). Thus, apart from their endemic faunas the Greater and Lesser Antilles, which are only a short distance away from the continent, have been populated by only a few pioneering winged forms, helped in their migration by the winds, or by riparian forms which arrived on rafts broken away from the banks. The same is true for the islands of Cap Vert, off Africa (composition of the fauna is given in Duranton *et al.* 1983). The Galapagos islands (Dirsh 1969), still further away and less favoured by wind direction, have acquired even fewer immigrants.

Nevertheless, these exchanges of species must not be neglected when discussing intercontinental dispersion even though they may be restricted in number and generally limited to a few ecological groups (i.e. the winged fauna of herbaceous habitats or geophiles), with good reproductive potential and generalist habits. However, for one case in present-day faunal distribution (see below the genus *Schistocerca*), they have only been effective over short distances.

The same is not true for other members of the Orthoptera. For example, the Gryllidae disperse easily and, probably for ecological reasons, colonize similar islands with a greater variety of forms than acridians (Otte and Rentz 1985; Otte *et al.* 1987; Otte 1988).

Communication routes

The ability of acridians to follow the biotopes to which they are adapted has allowed, and still allows, them to follow efficiently migration routes through continents offering an ecological continuum. This is evidenced by the composition of the present population of South America. However, there are again exceptions to the rule. For example, the North American Melanoplinae, while having colonized the Andes and prospered in the southern subtropical zone, are very far from having colonized the whole continent.

The Oedipodinae are geophilic forms which migrate easily and have great dispersal ability. Though these were established in North America in ancient times, only three genera are found in South America today.[1] One of the two main genera is distributed on the andean dry altiplano and along the coasts and desert areas of the Pacific as far as Chile but is hardly established east of the Andes. The second has followed the

---

[1] Except the genus *Sphingonotus*, which has a worldwide distribution and is known in the Caribbean and Galapagos Islands.

Atlantic coasts of Venezuela (*Heliastus*: Otte 1984). The third genus (*Lactista*) is apparently rare and is restricted to the territory of Roraima (unpublished observations) although climatic conditions favourable to its expansion in subtropical South America have frequently occurred in the recent past (forest regressions).

ANTIQUITY AND EVOLUTION OF ACRIDIANS—AFRICAN–AMERICAN RELATIONSHIPS

Knowledge of the age of extant phyla as well as accurate phyletic comparisons are essential for interpreting the observed relationships between Africa and America. We will thus summarize this point. For clarity, we will begin by eliminating the false relationships which are sometimes still reported.

Introduction—false relationships

The Cylindrachetidae are terricolous Orthoptera which are greatly modified and difficult to recognize. This group has a direct austral distribution (Australia/Patagonia–Chile) but no representatives in Africa. Such an austral relationship has not yet been described for acridians.

Although information provided by fossils is very useful to trace the general history of the group, it does not allow us to establish past continental relationships with the same precision as for the vertebrates. There are two reasons for this. First, now as in the past, family or suprafamily classification can generally be recognized only by dissection or by characters rarely preserved in fossils. Second, most of the known fossils come from northern continents.

Yet faunal similarities do exist at several levels between Africa and America, and various parallels have been drawn by different authors. These are often based on ecomorphogical convergences or on characters of a weak phyletic value. The list of postulated relationships now known to be erroneous could be long, but the most important recurring ones are the following.

Based on ecomorphological convergence

Such convergences are almost the general rule among acridians. All graminicolous species or all the geophilic species, for example, may acquire the same shape and the same cryptic and mimetic mechanisms regardless of their phyletic origin. This results in erroneous associations, as can be seen below.

1. The Euthymiae (Ramme 1929; Rehn 1938, 1944): an adaptable tree-dwelling group composed of the neotropical Romaleidae (Amedegnato 1977) and of genera of the Old World, particularly from Madagascar and Asia, now classified more or less satisfactorily among the Hemiacridinae (Dirsh 1961, 1962; Descamps and Wintrebert 1966; Dirsh and Descamps 1968).

2. The Hemiacridinae (Dirsh 1961): not clearly defined; on the basis of an apparent convergence of the stridulatory apparatus, this group combined two neotropical genera (*Aleuasini*: Amedegnato 1977) with an African group or worse, with the Hemiacrididae (Dirsh 1975) which are much more heterogeneous.

3. The ancient pantropical Catantopinae: now split into numerous families and subfamilies (Dirsh 1961; Amedegnato 1977).

4. The South American geophilic Ommexechidae: classified with the Pamphagidae and related African groups (Dirsh 1975).

5. The Episactinae of Central America placed together with the Miraculinae from Madagascar: both representative of bush-dwelling eumastacids found in the undergrowth (Descamps 1973*b*).

Based on the elytral venation or sometimes on some unknown characters

1. The grouping of Proscopiidae and Pneumoroidea, South-Central American-South Africa (Sharov 1968), which remains unexplained.[1]

[1] The relationship established between eumastacids, proscopids, and Trigonopterigidae (Dirsh 1975; Kevan 1977) is not essential to this discussion, but it remains unexplained and does not seem to be based on any factual observation.

2. The grouping of Trigonopterigidae (South Oriental region) and of Pneumoroidea South Africa, Central and North America) based on a modification of the elytral nervation (Sharov 1968).

Based on the preconceived idea that interrelationships among old groups of the southern continents must exist

The grouping of Morabinae (australian eumastacides) and Proscopiidae from South America (Rehn 1948; Blakith and Blakith 1968; Liana 1972; Mesa and Ferreira 1981). This grouping reflects only membership to a same large phyletic unit whose members bear 17 chromosomes. This is the primitive basic number common to all the eumastacids and proscopids of the world (White 1968; Hewitt 1979; Mesa and Ferreira 1981). The presence of cervical sclerites also characterizes all members of that phyletic unit.

In addition a number of true relationships exist at all levels but have been more or less obscured; what importance and what meaning should we give them?

No coherent and definite answer can be given without knowing the age of extant groups, their phyletic relationships and precise distribution, as well as their recent dispersals, which are often shown by clines of species that are more and more derived. Unfortunately, even a lack of data does not prevent some people from concluding that recent groups with recent distribution are relics of Pangea!

Fossil groups and extant groups

Sharov has proposed a synthesis of our knowledge about fossil Orthoptera, particularly caeliferous Orthoptera (Fig. 5.1a). Because of the problems mentioned above and to accomodate modern classification of the Caelifera, we have thought it necessary to modify his scheme (Fig. 5.1b).

With the exception of the Cylindrachetidae and Tridactylidae, which are only loosely related to the Caelifera, and with no close relationship to the Tetrigoidea, the main modifications concern:

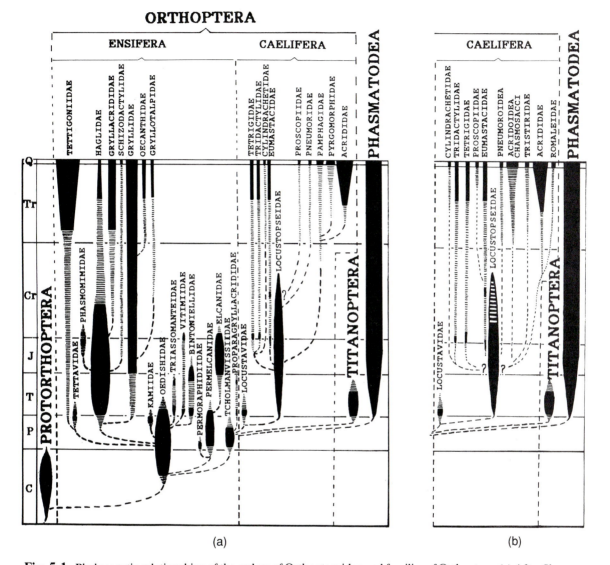

Fig. 5.1 Phylogenetic relationships of the orders of Orthopteroidea and families of Orthoptera. (a) After Sharov (1968, translated 1971); (b) modification and updating of the Caelifera.

1. the fusion of the eumasticids and proscopids, based on a greater number of characters and accepted by all authors, with the concomitant rejection of the Proscopiidae–Pneumoroidea relatedness.
2. The fusion of the Pamphagidae and Pyrgomorphidae into the Acridoidea Chasmosacci,

which also includes the South African Lentulidae, omitted by Sharov (plus some other groups or genera presently scattered through the classification of Old World acridians).
3. the inclusion and separation of the Tristiridae (again omitted by Sharov), related to the former, but isolated in South America and

intermediate with the Acridoidea *sensu stricto*.

4. the inclusion of the Romaleidae, specific to the New World, and their separation from Old World Acrididae.

The levels of divergence from the 'Locustopseidae' are roughly those given by Sharov and will be described later, along with the more ancient origin that we ascribe to the Pneumoroidea.

Sharov classifies the eumastacids within the Locustopsoidea. In fact, one does find among the Locustopseidae a mixture of fossils with typical eumastacoid and typical acridoid characteristics (mainly wings and elytra of locustopseid fossils). Knowing that high taxonomic level, such as Pneumoroidea and the Acridoidea, cannot be recognized in fossils, it is reasonable to think that if the eumastacids (and consequently the proscopids) belong to the Locustopsoidea, more likely to the Locustopseidae (see Sharov's diagram, Fig. 5.1a), then this family's living representatives are found among the other acridomorph groups. The same is true for the Haglidae and the Gryllidae within the Ensifera. Therefore, the Locustopseidae correspond to the general term 'acridians', or to the Acridomorpha of Dirsh, to encompass all the groups familiar to a specialist of the Orthoptera. One also notices that the 'Locustopseidae' disappear from the fossil layers with the appearance of modern faunas, which are identical to those of today and recognizable, as early as the Palaeogene.

Relationships between Africa or Old World/America in the Acridomorpha: facts and hypotheses

If we disregard the faunal similarities resulting from great ecomorphological convergences (savannah fauna, desert fauna, etc.) which will not be discussed here, the supposed phyletic relationships between Africa and America may be classified into three main categories:

1. South Africa/Madagascar/Central America distribution

2. extreme South America/South Africa distribution

3. relationships between Africa/America for a more recent fauna, at a lower taxonomic level.

These types of relationships respond differently and variably to analysis according to the groups studied. The most efficient way to examine these relationships and to understand the true nature of the faunal similarities observed between the two continents, while cross-checking phyletic similarities, history of the relationships, and hidden ecomorphological similarities, seems to be to focus on higher taxonomic levels. We will thus examine the relationships within superfamilies, which will be presented in chronological order.

Eumastacoidea (Fig. 5.2)

This is the most ancient and archaic group of the Acridomorpha and, for this reason, the relationships between Africa and America shown by the taxon are essential. Unfortunately, a clarification of the phyletic relationships within the superfamily is necessary prior to any discussion.

Eumastacoidea (comprising several families) and the Proscopiidae have always been thought to be related to each other, although the latter has been raised to the superfamily rank several times (Uvarov 1966; Descamps 1973*a*; and also Dirsh 1975) but always within the Order Eumastacoidea.

By studying the American fauna we have had to reconsider the relationships between these large groups present in the neotropical zone. We will give details of these relationships elsewhere and will only summarize, below, the organization of the Eumastacoidea.

Several authors have proposed classifications of the eumastacids which agree on some points but disagree on others. The most coherent system remains that proposed by Descamps (1973*b*), based on the subfamily system of Rehn (1948). He divides the eumastacids (the proscopids are excluded because of their divergence) into the Cryptophalli, Stenophalli, Disclerophalli, and Euphalli.

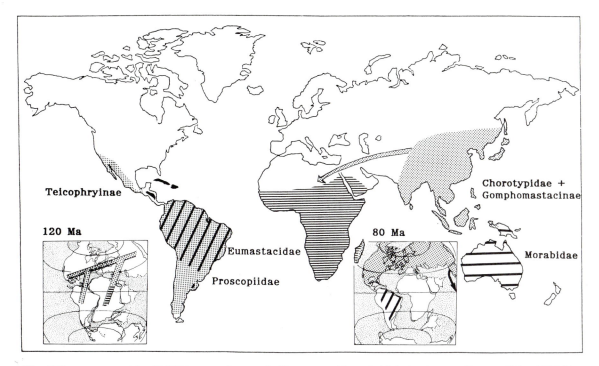

Fig. 5.2 Eumastacoidea. ▦Old group, Cryptophalli eumastacids and old derived group (Proscopiidae);☰African derived group, Disclerophalli eumastacids;◪☰More recent derived groups, Eumastacidae, Morabidae.

Observation of the structure of the male genitalia in these groups, as well as other characters allows us to ascertain the position of the American eumastacids. It also allows us to comment on these group divisions and, hence on the biogeographical relationships which can be deduced from their distribution.

The Disclerophalli correspond to a natural African assemblage. We can neglect the Indian Euphalli in this discussion.

The Stenophalli comprise the whole of the South American eumastacids and also constitute a natural assemblage if only American species are considered. By contrast, the Gomphomastacinae are also included which on the contrary are, connected by Dirsh with good reason, to the other Asian families. The Morabinae and Biroelleinae, which are also included with the Stenophalli, form a separate branch.

The concept of Stenophalli should therefore be abandoned. Two groups thus remain: the family Eumastacidae, which is American and homogeneous, and the family Morabidae (New Guinea, Australia). The close relationships of the Gomphomastacinae to other Asian groups, recognized only by Dirsh (1975), and to the Teicophryinae of Central America, suggest relating them to the Cryptophalli.

The Cryptophalli, considered to be the most ancient branch, is composed of the Asian groups. Under a different name (the Episactidae, Descamps 1973b), the Cryptophalli also include the Miraculinae of Madagascar, and the Teicophryinae and Episactinae of Central America.

In fact, the Miraculinae of Madagascar are only distantly related to the diverse Asian groups and bear strictly no relationship with the 'Episactidae' of Central America. The latter were found to be composed of two unrelated subfamilies: the Episactinae, with a reduced ectophallus, which are related to the whole of the South American subfamilies, where Dirsh has put them; and the Teicophryinae, which are very close to both the South American Proscopiidae and to the Asiatic Gomphomastacinae (the basic structure of the phallic complex is similar). These groups belong to the Cryptophalli (structure and function of the endophallus).

There is no doubt that the well diversified South American Proscopiidae constitute an ancient derivative of the most ancient group of the eumastacids (Cryptophalli)[1] (still present in Asia where it comprises a number of types) as are the African Disclerophalli, for example which are as divergent. The Teicophryinae are intermediate and probably survivors of the old Eumastacoidea fauna of the northern continent.

Apart from this, the South American population of the Eumastacidae, strictly restricted to the tropical region and not extending over the Patagonian shield, is homogeneous and more recent. However, it is not related to the wave of Plio–Pleistocene immigrants to South America (no relatives in the Old World). Likewise, the Australian eumastacids constitute a rather recent parallel line partly adapted to the Gramineae (Key 1959, 1976).

On the whole, the ancient phylum of Eumastacoidea, especially the Cryptophalli, appears to be a phylum of the northern continents having generated distinctive groups in the southern continents (Disclerophalli in Africa, Proscopiidae in America.[2] Similarily, a second movement, also

starting from the northern continents, has given rise to the South American and Australian faunae. However, some groups of differing antiquity, whose phylogeny remains to be analysed, may exist in the fauna of South America.

Since one must remove both the relationship of Africa and America based on the distribution of the Episactidae (Descamps 1973*b*) and the relationships of the Proscopiidae with the Morabinae (see above), it appears that no direct and presently discernible relationships between the African and American faunas, or even between the faunas of the southern continent, exist for the Eumastacoidea. However, one exception is the extension of the Asian Chorotypidae (Cryptophalli) into the North American continent, where they are known by an Oligocene fossil (Lewis 1976), and into Africa where a similar genus, *Hemerianthus*, is known.

Pneumoroidea (*Fig. 5.3*)

The Pneumoroidea which cannot presently be related to any other acridian type, show a clear African-American distribution at the superfamily level. However, the basic structure of the genitalia and the unique structure of the stridulatory apparatus are the only characteristics common to the three known families, which are otherwise extremely different. Note that the Tanaoceridae do diverge in structure of the genitalia. Their very restricted distribution and weak dispersal ability indicate that they are true relics of an ancient fauna which was probably more widespread. The present distribution of the Pneumoridae in South Africa, of the Xyronotidae in isolated areas of low altitude Central American forests (Chiapas, Vera Cruz), and of the Tanaoceridae in Baja California and in the South-west of the USA (arid zone) cannot be interpreted with any certainty, although it may truly reflect the existence of an ancient acridomorph fauna common to both continents.

Acridoidea Chasmosacci and Tristiridae (*Fig. 5.3*)

We have put several systematic groups on the same distribution map to show clearly the past

[1] The Proscopiidae and the Asian groups also share a common ancestral character, common to early Orthoptera, and which disappears many times (i.e. the fundamental configuration of the apical spurs of the posterior tibias, three in spurs on each side). Moreover, some genera of proscopids show several spurs on the anterior tibial face, which are not present in any other modern acridomorph phylum. This suggests an ancient splitting of this group.

[2] To simplify, we have shown these first vicariance events, which were probably not contemporary, on the map of the continents of 120 Ma ago (Smith 1981); however, we cannot yet provide reliable details on the time of formation of the Proscopiidae in South America and of Disclerophalli in Africa.

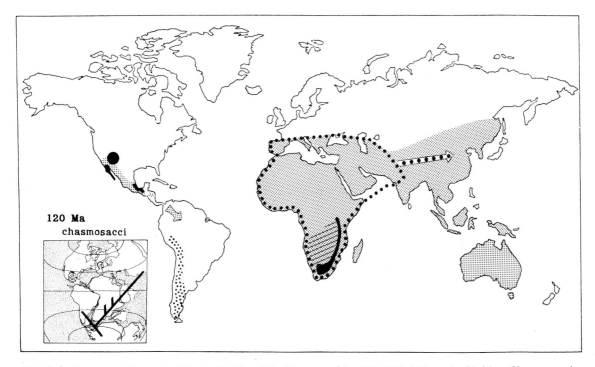

Fig. 5.3 Pneumoroidea and old Acridoidea. ■, Pneumoroidea; ▦, Tristiridae; Acridoidea Chasmosacci: •••, Pamphagidae; ▦, Pyrgomorphidae; ▨, Lentulidae.

relationships between the two main southern continents.[1] The Pamphagidae and Pyrgomorphidae have long ago been included under this name (Roberts 1941); they are two derived families. By contrast, the Tristiridae and Lentulidae are not usually added to this phyletic unit, although at least one of these groups is closely related to it.

The Tristiridae are a primitive family, with characteristics intermediate with those of the other great group of Acridoidea (A. Cryptosacci) (Eades 1962; Eades and Kevan 1974; Amedegnato 1976, 1977).

The phallic complex of the Lentulidae shows the same basic structure as that of the Pyrgomorphidae (ventral sclerification of the sperm-atophore sac and structure of the ectophallus);[2] as previously suggested (Amedegnato 1976), we will thus consider them parts of the same unit.

The Tristiridae are endemic to South America. They are characteristic of Patagonia, and northward through the Andes. The few existing genera are divided into various types (Amedegnato 1977; Cigliano 1989) and show some alate and apterous geophilic forms, as well as forms which inhabit the leaf litter, proving the ancient adaptation of the group to Cenozoic forest undergrowth. This diversified faunal group, whose weak dispersal capability did not allow it to colonize the rest of South America, seems relatively recent on the South American continent and probably came from the Antarctic.

---

[1] This is not valid for the Pneumoroidea, also mentioned on this map only because of their antiquity.

[2] We would need to link to the Lentulidae some more or less recognized and classified related elements.

The Pamphagidae are fundamentally endemic to the African plate, although they could have spread to the south of Europe and Asia.

The Lentulidae are ecologically highly varied in South and tropical Africa. With the Pamphagidae, they must have constituted the ancient African acridian fauna. They are unknown in the northern continents.

The Pyrgomorphidae are closely related to the Lentulidae and seemingly recent. They appear to be geographically vicariant, originating in Asia, later returning to Africa and extending, to a lesser degree, into America.

Thus, this faunal group with its four main families seems to characterize well the southern continents and may have originated there. Nevertheless, the Tristiridae, which may represent today's remnants of the antarctic fauna, have numerous intermediate characters (genitalia and external morphology) relative to the Acridoidea Cryptosacci. They may therefore be very old and their Patagonian distribution may be a relict.

Acroidoidea Cryptosacci (Fig. 5.4)

These are the modern acridians. Divided into a multitude of systematic groups, they are distributed throughout the world, where they occupy every environment. Three main families exist: Romaleidae and Ommexechidae of the New World, and the Acrididae, characteristic of the Old World. The latter show the greatest number of the faunal similarities between Africa and America.

The Romaleidae radiated early enough in South America to constitute the basic fauna of that continent. They undertook two successive northward expansions. The most recent involves the tropical genera presently distributed on both

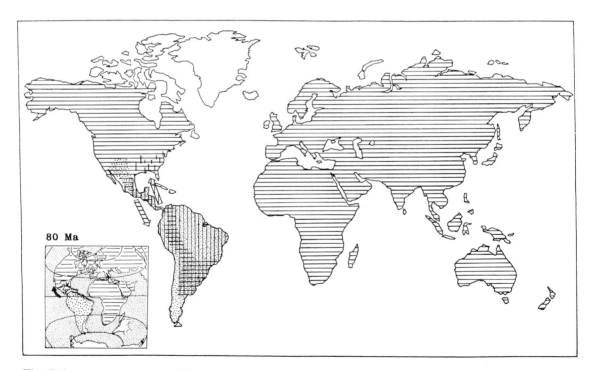

Fig. 5.4 Modern Acridoidea. ▨ Romaleidae, old neotropical vicariance within the Acridoidea, ▥ Acrididae, South American subfamilies (Cenozoic vicariance), ▤ Acrididae, distribution of the fauna originating from the Old World.

sides of the Mexican Altiplano. The other, more ancient extension gave rise to diversified endemic forms in the western USA and on the Mexican Altiplano. These species, morphologically convergent with the Oedipodinae and Pamphagidae of the Old World, are now integrated into the local ecosystem which include mostly nearctic fauna.

The Romaleidae, which may be the most primitive of the true acridians, may have resulted from an ancient vicariance event separating South America from North America and the Old World. The Ommexechidae, which seem closely related but very derived, are from the south of the continent. This group shows no relationship between Africa and America other than that to the common Acridoidea ancestor.

The Acrididae, which constitute the bulk of the modern Acridoidea fauna and include the most advanced forms, are spread throughout the world. They originated from the Old World and formed the vicariant sister-group of the Romaleidae. However, in America they are clearly divided into a nearctic and neotropical fauna. The later, typically South American, evolved during the Tertiary occuping all environments and frequently replacing the Romaleidae (palm trees) or occupying relatively free niches (Gramineae). This fauna gave rise to phyla which are equivalent, but different, to those of the Old World and of the nearctic region, and which show the two distinct important ecological adaptations (living on shrub- or tree-like dicotyledons or on monocotyledons, mostly Gramineae).

As for other animals during the last continental connection, the southern orthopteran forms migrated northward only slightly while nearctic forms colonized South America massively.

As a result, the native thamnophilous forms, which are adapted to dicotyledons, encountered in the andean and subtropical regions to the south of the amazonian block, the rapidly speciating Melanoplinae (which are closely related, if not equivalent, to the Podisminae of Asia and the true Catantopinae of Africa). In forested regions, the Proctolabinae, a related tree dwelling group

speciated similarly quickly by colonizing secondary forest growth. The expansion of this subfamily, recently arrived in South America, was influenced by physical barriers to dispersal. Indeed, dispersal is largely limited to West and South Amazonian centres, the latter being more ancient than the former.

The ancient South American graminicolous forms had developed more advanced adaptations than the immigrating Acridinae and Gomphocerinae (African and Asian), which are mostly pioneers. These adaptations (especially endophytic egg laying) protected them in humid environments, not yet colonized by the newcomers. Similarly, palm trees hosted a specific fauna (Copiocerini) which the newcomers, having no equivalent forms, were unable to supplant.

So, after the first and ancient vicariance separating the Acrididae and Romaleidae, a second Cenozoic vicariance occurred, splitting Old World acridians from their South American counterparts, as took place in the mammals. this vicariance is now followed by the interaction of the two faunas in the same continent, the new fauna showing most of the close African–American relationships (five subfamilies). However, the closest generic relatives are the Asian, including fossils dating back to the beginning of the Cenozoic, especially from the Oligocene, in North America (Lewis 1976). A portion of the population of South America which seems either recent (but certainly not Plio–Pleistocene) or anomalous, could have had as ancestors relatively mobile migrants from the northern continents of that period (Pygomorphidae, Copiocerinae other than Copiocerini) (Amedegnato 1977).

The Schistocerca case (Fig. 5.5)

One of the most spectacular examples of the African-South American relationships, which gave rise to many speculations, is the colonization of America by the genus *Schistocerca*. Common to Africa and America and belonging to a group well diversified in the Old World, this genus is represented in Africa by a single species, while in

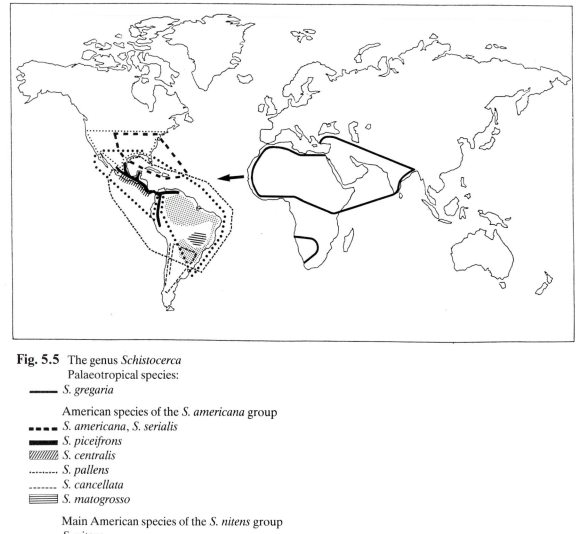

Fig. 5.5 The genus *Schistocerca*
     Palaeotropical species:
——— *S. gregaria*

American species of the *S. americana* group
▬ ▬ ▬ *S. americana, S. serialis*
▬▬▬▬ *S. piceifrons*
▨▨▨ *S. centralis*
·····―· *S. pallens*
------- *S. cancellata*
≡≡≡ *S. matogrosso*

Main American species of the *S. nitens* group
······· *S. nitens*
▨▨▨ *S. flavofasciata*

·········· Other North American species

America it diversified into many species.[1] It is also the most divergent taxon of the Cyrtacanthacridinae. This divergence and the abundance of species led some scientists to suggest America as

[1] The genus *Halmenus* Scudder 1893 from the Galapagos is an immediate derivative of *Schistocerca* (insular brachypterism).

the cradle of this genus; but most saw the large number of species as indicative of radiation by a recent colonizing genus into a free niche in a new continent. The transAtlantic crossing by the gregarious African form, at the time of the recent gregarious phase of the African *Schistocerca*, probably ruled in favour of the latter. However, a definitive and exclusive conclusion, favouring just one migratory transatlantic route, may not be appropriate.

Indeed, *Schistocera* is represented by several groups of species in America, the main ones being *S. americana* (Harvey 1981) and *S. nitens*. The *S. nitens* group is probably older than other groups since it is well divergent and it does not hybridize with either the *S. americana* group or with *S. gregaria*. In the latter respect, it is similar to the other nearctic species. *S. pallens*, the nucleus of the *S. americana* group and *S. gregaria* all hybridize to a greater or lesser extent (without producing fertile offspring: Jago *et al.* 1979); they are much more closely related and probably result from a more recent extension. It is likely that *S. gregaria*'s only direct relatives are species of the *S. americana* group. The existence of a divergent species in Florida and of a Cuban endemic genus of Cyrtacanthacridinae (in which the male is undescribed)[1] confirms that the group must have been more diversified in North America, with nearctic species of the genus being part of this large taxon. The group's diversity was later greatly reduced by glaciations. From then on, some of the species or the genus itself could belong, like the Acridinae, Gomphocerinae, Oedipodinae, and Melanoplinae, with other Cyrtacanthacridinae to the Cenozoic North Atlantic fauna.

CONCLUSIONS

The examination of present distributions linked with that of the modes and patterns of speciation and diversification allows us, to a certain extent, to postulate the relationships between African and American fauna in the acridian Orthoptera. However, it may not be possible to look very far into the past in any other way than by posing uncertain hypotheses. Moreover, although the warm climate of most of the Palaeozoic favoured the global development of the group, subsequent

cooling during the Tertiary which culminated in the last ice age, caused many faunal modifications and extinctions. The composition of the South American fauna was thus extensively altered (Amedegnato 1977). Similarly, the present tree-dwelling African fauna appears incredibly poor, although it must have been well diversified in the past, as demonstrated by 'living fossils' still present as well as by the tree-dwelling fauna of Madagascar. Interesting clues have probably disappeared, and this contributes greatly the incompleteness of the data.

In spite of this, the fossils dating back to the Oligocene of North America and the Miocene of East Africa show that, at the generic level, the Cenozoic faunas on the great continental blocks were already identical to the present fauna, except probably in distribution (disappearance of the North American Chorotypidae). Therefore, by the end of the Mesozoic, South America and Africa were likely already inhabited by their ancient groups, which were older than the fauna of the Tertiary and which probably developed simultaneously with the angiosperms.[2] We may thus put forward hypotheses, based on these great phyla which are still represented in the modern populations. Indeed, we saw that phylogeny at the superfamily level can reveal the existence of ancient relationships, although these relationships appeared very weak. One must not forget that long before their final separation by an ocean, Africa and America were separated by ancient systems of rifts, with volcanic eruptions, and by seas of varying depth and permanence. These barriers must have influenced the evolution of high-level taxonomic groups.

Moreover, the most archaic groups of the Acridomorpha (Eumastacidae and Proscopiidae) are

---

[1] *Nichelius* Bolivar 1888.

[2] The argument which links the antiquity of eumastacoids to the antiquity of their diet (some species eating ferns, Blakith 1973; Descamps 1973) does not seem necessary to us and is probably not correct. Indeed, most species eat dicotyledonous angiosperms. Moreover, some species of modern acridians in pteridophyte-dominated environments have adapted to this diet, for example *Hylopedetes* (Rowell 1978). Indeed, this argument was immediately refuted (Key 1976; see also Balick *et al.*1978; Cooper-Driver 1978).

unlikely to have split because of the break-up of Pangea (Blakith 1973). Given the antiquity of this group, as shown by many very primitive morphological characters, this could indicate that, probably as for other Locustopseidae, these groups did not originate on the southern continents, or that the southern fauna disappeared without leaving any traces. By contrast, the more recent groups (Pneumoroidea?, Acridoidea) have had representatives on the northern as well as on the southern continents, resulting in distant faunal relationships between the two southern continents.

As one tries to understand the history of younger group, it becomes clear that the great faunal similarities between Africa and America have arisen via migrations from the northern continents, except for a few events of long distance dispersal that have generated no distinct groups.[1]

On the whole, the faunas of the northern continents contrast by their homogeneity with the disparity and derived characters of the fauna of the southern continents. For the acridids, this reflects influence of palaeogeography on the evolution of the group. In addition, northern faunas have long been, and still are, more successful in colonizing the southern continent than southern faunas invading the north.

In many ways, the biogeography of acridians and the development of their faunas are similar to those of recent vertebrates. However, in contrast to vertebrates, ancient acridian phyla have survived and still play a very important role in the population structure of great terrestrial ecosystems.

REFERENCES

Amedegnato, C. (1976). Structure et évolution des génitalia chez les Acrididae et familles apparentées. *Acrida*, **5**, 1–16.

Amedegnato, C. (1977). Etude des Acridoidea centre et sud americains (Catantopinae sensu lato). Anatomie des génitalia, classification, répartition, phylogénie. Unpublished Thesis, University of Paris VI, 385 pp.

Amedegnato, C. and Descamps, M. (1982). Dispersal centers of the Amazonian Acridids. *Acta Amazonica*, **12**(1), 155–65.

Balick, M. J., Furth, D. G. and Cooper-Driver, G. (1978). Biochemical and evolutionary aspects of arthropod predation on ferns. *Oecologia*, **35**, 55–89.

Blakith, R. E. (1973). Clues to the Mesozoic evolution of the Eumastacidae. *Acrida*, **2**, V–XV.

Blakith, R. E. and Blakith, R. M. (1968). A numerical taxonomy of Orthopteroid insects. *Australian Journal of Zoology*, **16**, 111–31.

Cigliano, M. M. (1989). Revision sistematica de la familia Tristiridae (Orthoptera, Acvidoidea). *Bolotin de la Sociedad de Biologia de Conception, Chile*, **60**, 51–110.

Cooper-Driver, G. A. (1978). Insect fern association. *Entomologia experimentalis et applicata*, **24**, 310–16.

Descamps, M. (1973*a*). Notes préliminaires sur les génitalia de Proscopoidea (Orthoptera, Acridomorpha). *Acrida*, **2**, 77–95.

Descamps, M. (1973*b*). Révision des Eumastacoidea (Orthoptera) aux échelons des familles et des sous-familles (génitalia, répartition, phylogénie). *Acrida*, **2**, 161–98.

Descamps, M. (1977). Monographie des Thericleidae (Orthoptera, Acridomorpha, Eumastocoidea). *Annales du Musée royal de l'Afrique centrale, Sciences biologique*, **216**, 475 pp.

Decamps, M. (1984). Revue préliminaire de la tribu des Copiocerini (Orthoptera, Acrididae). *Mémoires du Muséum National d'Histoire Naturelle* Nouvelle Séries, Série A, Zoologie, **130**, 72 pp.

Descamps, M. and Wintrebert, D. (1966). Pyrgomorphidae et Acrididae de Madagascar; observations biologiques et diagnoses. *Eos*, 250 pp.

Dirsh, V. M. (1961). A preliminary revision of the families and subfamilies of Acridoidea (Orthoptera, Insecta). *Bulletin of the British Museum of Natural History (Entolomology)*, **10**, 351–419.

Dirsh, V. M. (1962). The Acridoidea of Madagascar, I, Acrididae (except Acridinae). *Bulletin of the British Museum (Natural History) Entomology*, **12**(6), 25–193.

Dirsh, V. M. (1966). Acridoidea of Angola, I, II, Museu do Dundo, Subsidios para o estudo da biologia na

---

[1] Given the similarity observed between Mid-Cenozoic and modern faunas, the transatlantic crossing of the genus *Schistocerca* may occurred over a long period (Kevan 1989).

Lunda. *Publicacoes culturais da Companhia de diamantes de Angola*, **74**, 11–305, 307–527.

Dirsh, V. M. (1969). Acridoidea of the Galapagos Islands (Orthoptera). *Bulletin of the British Museum (Natural History), Entomology*, **23**(2), 51 pp.

Dirsh, V. M. (1975). *Classification of the acridomorphoid insects*, 171 pp. E. W. Classey, Faringdon, Oxon.

Dirsh, W. M. and Descamps, M. (1968). Insectes Orthoptères Acroidoidea, Pyrgomorphidae et Acrididae. *Faune de Madagascar*, **26**, 312 pp. ORSTOM-CNRS, Paris.

Duranton, J. F., Launois, M., Launois-Luong, M. H., and Lecoq, M. (1983). Contribution à l'inventaire faunistique des Acridiens de l'Archipel du Cap Vert (Orth.). *Bulletin de la Société entomologique de France*, **88**(3,4), 197–224.

Eades, D. C. (1962). Phallic structures, relationships and components of the Dericorythinae (Orthoptera, Acrididae). *Notulae Naturae*, **354**, 1–9.

Eades, D. C. and Kevan, D. K. McE. (1974). The phallic musculature of Pyrgomorphidae, with particular reference to *Atractomorpha sinensis* Bolivar, and notes on the family Tristiridae and the subfamily Pyrgacridinae, nov. (Orthoptera, Acridoidea). *Acrida*, **3**, 247–65.

Harvey, A. W. (1981). A reclassification of the *Schistocerca americana* complex (Orthoptera : Acrididae). *Acrida*, **10**, 61–77.

Hewitt, G. M. (1979). Orthopera: Grasshoppers and Crickets. *Animal Cytogenetics*, **3**, Insecta 1, IV and 170 pp.

Jago, N. D., Antoniou, A., and Scott, P. (1979). Laboratory evidence showing the separate species status of *Schistocerca gregaria*, *americana* and *cancellata* (Acrididae, Cyrtacanthacridiane). *Systematic Entomology*, **4**, 133–42.

Kevan, D. K. McE. (1977). Suprafamilial classification of 'Orthopteroid' and related insects: a draft scheme for discussion and consideration. In *The higher classification of the orthopteroid insects* (ed. D. K. McE. Kevan), *Lyman Entomological Museum and Research Laboratory Memoir*, **4** (Special publication 12), appendix 1–26.

Kevan, D. K. McE. (1989). Transatlantic travellers. *Antenna*, **13**, 12–16.

Key, K. H. L. (1959). The ecology and biogeography of australian grasshoppers and locusts. In *Biogeography and ecology in Australia*. Series *Monographiae biologicae*, **8**, 192–210.

Key, K. H. L. (1976). A generic and suprageneric classification of the Morabinae (Orthoptera: Eumastacidae), with description of the Type species and a bibliography of the subfamily. *Australian Journal of Zoology*, supplementary series, suppl. 37, 1–185.

Lewis, S. E. (1976). A new specimen of fossil grasshopper (Orthoptera: Caelifera) from the Ruby River Basin (Oligocene) of southwestern Montana. *Annals of the Entomological Society of America*, **69**, 120.

Liana, A. (1972). Etudes sur les Proscopiidae (Orthoptera). *Annales Zoologici Varzawa*, **29**(13), 381–459.

Mesa, A. and Ferreira, A. (1981). Have the Australian Morabinae and the Neotropical Proscopiidae evolved from a common ancestor? A cytological point of view (Orthoptera, Caelifera, Eumastacoidea). *Acrida*, **10**, 206–17.

Müller, P. (1973). *The dispersal centers of terrestrial vertebrates in the neotropical realm*, a study in the evolution of the neotropical biota in its native landscapes. *Biogeographica*, **2**, 244 pp. Dr W. Junk, The Hague.

Otte, D. (1981). *The North American Grasshoppers*. I. Acrididae, Gomphocerinae and Acridinae. ix and 275 pp. Harvard University Press.

Otte, D. (1984). *The North American Grasshoppers*. II. Acrididae, Oedipodinae. viii and 366 pp. Harvard University Press.

Otte, D. (1988). Bark Crickets of the Western Pacific Region (Gryllidae: Pteroplistinae). *Proceedings of the Academy of Natural Sciences of Philadelphia*, **140**, 281–334.

Otte, D. and Rentz, D. C. F. (1985). The Crickets of Lord Howe and Norfolk Islands (Orthoptera, Gryllidae). *Proceedings of the Academy of Natural Sciences of Philadelphia*, **137**, 79–101.

Otte, D., Alexander, R. D., and Cade, H. (1987). The Cricket of New Caledonia (Gryllidae). *Monographs—Academy of Natural Sciences of Philadelphia*, **139**, 375–457.

Ramme, H. (1929). Afrikanische Acrididae. Revisionen und Beschreibungen wenig bekannter und neuer Gattungen und Arten. *Mitteilungen aus dem Zoologishes Museum in Berlin*, **15**(2), 247–492, 16 pl.

Rehn, J. A. G. (1938). A revision of the neotropical Euthymiae (Cyrtacanthacridinae). *Proceedings of the Academy of Natural Sciences of Philadelphia*, **90**, 41–102.

Rehn, J. A. G. (1944). A review of the Old World

Euthymiae (Orthoptera, Acrididae, Cyrtacanthacridinae). *Proceedings of the Academy of Natural Sciences of Philadelphia*, **96**, 1–135.

Rehn, J. A. G. (1948). The acridoid family Eumastacidae (Orthoptera) a review of our knowledge of its components, features and systematics, with a suggested new classification of its major groups. *Proceedings of the Academy of Natural Sciences of Philadelphia*, **100**, 77–139.

Roberts, H. R. (1941). A comparative study of the subfamilies of Acrididae (Orthoptera) primarily on the basis of their phallic structures. *Proceedings of the Academy of Natural Sciences of Philadelphia*, **93**, 201–46.

Rowell, C. H. F. (1978). Food plant specificity in neotropical rain forest acridids. *Entomologia experimentalis et applicata*, **24**, 451–62.

Scudder, S. H. (1893). The Orthoptera of the Galapagos Islands. *Bulletin of the Museum of Comparative Zoology at Harvard College*, **25**(1).

Sharov, A. G. (1968). Filogeniya ortopteroidnykh nasekomykh. *Transactions of the Institute of Paleontology*, 118 Moscou. (*Phylogeny of the Orthopteroidea* (trans. J. Salkind), 1971, Israel program for scientific translations Ltd, vi and 251 pp.)

Simpson, B. B. and Haffer, J. (1978). Speciation patterns in the Amazonian Forest biota. *Annual Review of Ecology and Systematics*, **9**, 497–518.

Smith, A. G. (1981). Phanerozoic equal-area maps. *Geologische Rundschau*, **70**(1), 91–127.

Uvarov, B. (1966). *Grasshoppers and Locusts*. I. Anatomy, physiology, development, phase polymorphism, introduction to taxonomy, xi and 481 pp. Cambridge University Press.

White, M. J. D. (1968). Karyotypes and nuclear size in the spermatogenesis of grasshoppers belonging to the subfamilies Gomphomastacinae, Chininae and Biroellinae (Orthoptera, Eumastacidae). *Caryologia*, **21**(2), 167–79.

FURTHER READING

Dirsh, V. M. (1956). The phallic complex in Acridoides (Orthoptera) in relation to taxonomy. *Transactions of the Royal Entomological Society of London*, **108**, 223–356.

Dirsh, V. M. (1974). Genus *Schistocerca* (Acridomorpha, Insecta), series Entologica (ed. E. Schimit-schek, Göttingen) 10, VII and 238 pp. Dr W. Junk, The Hague.

Donnelly, T. H. (1988). Geological constraints on Caribbean Biogeography. In *Zoogeography of Caribbean insects* (ed. J. k. Liebherr), pp. 15–37. Cornell University Press, London.

Haffer, J. (1978). Distribution of Amazon forest Birds. *Bonner Zoologische Betrage*, **29**(1,3), 38–78.

Haffer, J. (1979). Quaternary Biogeography of Tropical Lowlands South America. In *The south american Herpetefauna: its origin, evolution and dispersal* (ed. H. E. Duellman). *University of Kansas Museum of Natural History Monograph*, **7**, 107–40.

Haffer, J. (1982). General aspects of the refuge theory. In *Biological diversification in the tropics* (ed. G. T. Prance), pp. 6–24. Columbia University Press, New York.

Hammen, T. van der (1974). The Pleistocene changes of vegetation and climate in tropical south America. *Journal of Biogeography*, **1**, 3–26.

Hammen, T. van der (1982). Paleoecology of Tropical South America. In *Biological diversification in the tropics* (ed. G. T. Prance), pp. 60–6. Columbia University Press, New York.

Hoffstetter, R. (1971). Le peuplement mammalien de l'Amérique du sud. Rôle des continents austraux comme centres d'origine, de diversification et de dispersion pour certains groupes mammaliens. *Anais de Academia Brasileira de Ciencias*, **43** (suppl., 125–44.

Kevan, D. K. McE. and Wighton, D. C. (1981). Paleocene Orthopteroids from south-central Alberta, Canada. *Canadian Journal of Earth Sciences*, **18**, 12, 1824–37.

Kevan, D. K. McE. and Wighton, D. C. (1983). Further observations on North American Tertiary Orthopteroids (Insecta: Grylloptera). *Canadian Journal of Earth Sciences*, **20**, 2, 217–24.

Key, K. H. L. (1972). A revision of the Psednurini (Orthoptera: Pyrgomorphidae). *Australian Journal of Zoology*, supplementary series. Suppl. 14, 1–72.

Lewis, S. E. (1974). Four specimens of fossil grasshoppers (Orthoptera: Caelifera) from the Ruby River Basin (Oligocene) of southwestern Montana. *Annals of the Entomological Society of America*, **67**, 523–5.

Malfait, B. T. and Dinkelman, M. G. (1972). Circum-caribbean tectonic and igneous activity and the evolution of the caribbean plate. *Geological Society of America Bulletin*, **83**, 251–72.

Rage, J. C. (1988). Histoire paléogéographique des

Vertébrés terrestres depuis la fin du Paléozoïque, principaux évènements. *Comptes rendus de la Société de Biogéographie*, **64**(1), 3–17.

Ragge, D. R. (1955). The wing venation of the Orthoptera Saltatoria with notes on Dictyopteran wing venation. *Bulletin of the British Museum (Natural History)*. Entomology, 159 pp.

Riccardi, A. C. (1987). Cretaceous Paleography of Southern South America. *Paleogeography, Paleoclimatology, Paleoecology*, **59**, 169–95. Elsevier Science Publishers, B. V. Amsterdam.

Ritchie, J. M. (1989). Miocene Grasshoppers from East Africa. *5th International meeting of the Orthopterists' Society*. Valsain (Segovia), Spain. July 17th–20th.

Simpson, G. G. (1969). South american mammals. In *Biogeography and ecology in South America* (ed. E. J. Fittkau, J. Illies, H. Klinge, G. R. Schwalbe, and H. Sioli), *Monographiae Biologicae*, **19**, 879–909.

Sykes, L. R., McCann, W. R., and Kafka, A. L. (1982). Motion of Caribbean plate during last 7 million years and implications for earlier Cenozoic movements. *Journal of Geophysical Research*, **87**, B.13, 10656–76.

Uvarov, B. P. (1937). The South American acridids with old world effinities (Orthoptera). *Revista de la Sociedad Entomologica de Argentina*, **9**, 3–5.

Uvarov, B. (1977). *Grasshoppers and Locusts*. II, Behaviour, ecology, biogeography, population dynamics, ix and 613 pp. Centre for Overseas Pest Research.

Zeuner, F. E. (1939). *Fossil Orthoptera Ensifera*. British Museum (Natural History), London, 2 vols., 321 pp., 80 pl.

Zeuner, F. E. (1941). The Fossil Acrididae (Orthoptera, Saltatorial). Part 1. Cantantopinae. *Annals and Magazine of Natural History*, 11th. series, **8**, 510–22.

Zeuner, F. E. (1942*a*). The Fossil Acrididae (Orthoptera, Saltatoria). Part 2. Oedipodinae. *Annals and Magazine of Natural History*, 11th series, **9**, 128–34.

Zeuner, F. E. (1942*b*). The Fossil Acrididae (Orthoptera, Saltatoria). Part 3. Acridinae. *Annals and Magazine of Natural History*, 11th series, **9**, 304–14.

Zeuner, F. E. (1942*c*). The Locustopseidae and the Phylogeny of the Acridoidea (Orthoptera). *Proceedings of the Royal Entomological Society of London*, (B), **11**(1), 1–19.

6 'GRUBE MESSEL' AND AFRICAN–SOUTH AMERICAN FAUNAL CONNECTIONS

Gerhard Storch

INTRODUCTION

The fossil record of the European Early Tertiary includes various terrestrial vertebrates which are most unexpected and of an 'exotic' appearance palaeobiogeographically. Outside Europe, they are known in the late Cretaceous and the Cenozoic of West Gondwanaland, in particular of South America, but they are unknown in the Late Cretaceous and the Early Tertiary of North America. The presence of these vertebrates thus suggests former land connections between Africa and South America and a crossing of the marine Tethys barrier between Africa and Europe by terrestrial routes. European palaeofaunas might, therefore, contribute to an understanding of African–South American faunal relationships. These groups are: ceratophryine anurans (Rage 1981), ziphodont mesosuchian crocodiles (Buffetaut 1988), struthioniform birds (ratites) (Houde 1986; Peters 1988a), phorusrhacid birds (Mourer-Chauviré 1981; Peters 1987), and myrmecophagid edentates (anteaters) (Storch 1981). Cariamid birds (seriemas) (Mourer-Chauviré 1983; Peters 1988b, and didelphid marsupials can be added (Crochet and Sigé 1983; Von Koenigswald and Storch 1988), but unlike the preceding taxa they are also known as Early Tertiary fossils in North America.

This vertebrate assemblage in total is best represented by fossil findings from 'Grube Messel' near Frankfurt, Germany. From this site two of the above groups—phorusrhacids and anteaters—are renowned as the biochronologically earliest records and by their extraordinary degree of preservation. Ratite and didelphid fossils, too, are remarkably well-preserved. The Messel seriemas are represented by distal foot bones and a skull, and the mesosuchian by a fragmentary snout and lower jaw; only ceratophryines have not, as yet, been discovered from this locality.

THE MESSEL SITE

The former opencast oilshale mine 'Grube Messel' is renowned world-wide for the diversity and completeness of its vertebrate, insect, and plant fossils (for synopsis see Schaal and Ziegler 1988). The fossiliferous sediments were deposited about 50 my ago into a small freshwater lake of tectonic origin (Franzen 1990). For some hundred thousands of years the lake acted as a depository for a rich flora and fauna flourishing under tropical to subtropical climatic conditions. Complete vertebrate carcasses slowly sank to the lake bottom where anoxic conditions and very weak water currents prevented them from being disintegrated and dispersed. Largely complete and articulated skeletons became entombed within the argillaceous sediments. Many fossils even have the outline of the soft body and the gut contents preserved. The extraordinary preservation thus permits safe taxonomic assignments and quite often reliable palaeobiological reconstructions, which is, at least, helpful for drawing any palaeobiogeographical conclusions.

Messel is early Middle Eocene in age. The biochronological assignment is essentially based on mammals. Accordingly, the fauna represents the European Land Mammal Age of the Early Geiseltalian (= Early Lutetian, Franzen and Haubold 1986) which corresponds with the

Mammalian Reference Level MP 11 (Schmidt-Kittler 1987).

The palaeobiogeographical history of the Messel vertebrates is complex both chronologically and geographically (Storch 1984, 1986; Storch and Schaarschmidt 1988). Europe was fully surrounded by oceans during the times when the Messel fossils were deposited and this isolation is reflected by a high degree of endemism, specifically among mammals. The biogeographical roots of the animals can be reconstructed in many cases and now, despite European Middle Eocene provincialism, the palaeogeographic scenario has to be expanded considerably, comprising all continents except Antarctica and Australia. We have to examine each taxon, in detail, in order to trace its geographical origins. The present paper focuses on land vertebrates with West Gondwanan (African-South American) affinities.

GONDWANAN AFFINITIES OF MESSEL VERTEBRATES

The mesosuchian crocodile *Bergisuchus dietrichbergi*

Bergisuchus is the only mesosuchian found in Messel, whereas eusuchians are much more abundant and diversified, five genera being known to date. It is characterized by a dorsally narrow and deep snout with steeply sloping sides, by a markedly enlarged fourth lower tooth, by teeth being slightly compressed transversally and partly serrated, and by the presence of an anteorbital fenestra. This mesosuchian was originally assigned to aff. *Sebecus* ? within the suborder Sebecosuchia (Berg 1966) and subsequently formally named by Kuhn (1968). Since then it was either referred to the sebecosuchian Baurusuchidae or it was regarded as a family, *incertae sedis*, within sebecosuchians.

Sebecosuchia most likely originated in South America in the upper Cretaceous after the continent had become separated from Africa. The disjunct distribution was explained by dispersal via a Central American land connection to North America during the Upper Cretaceous, by way of northern Atlantic land corridors to Europe during the Early Palaeogene (Buffetaut 1980). This interpretation was somewhat altered by the discovery of a ziphodont mesosuchian in the late Early Eocene locality of El Kohol in southwestern Algeria (Buffetaut 1982). The finding provided the first evidence to support an alternative hypothesis of faunal connections between South America and Europe, i.e. that the animals may have reached Europe from Africa instead of North America, implying the crossing of the proto-South-Atlantic and the Tethys.

Recently Buffetaut (1988) redescribed the Messel specimen and proposed a new taxonomic setting. Close comparisons with a number of mesosuchians from South America, Africa, and Europe revealed that *Bergisuchus* does not exhibit close similarities with the ziphodont forms from the South American Tertiary, but seems to be more closely related to the Late Cretaceous family Trematochampsidae from Africa and South America. Accordingly, he assigned the Messel taxon to the Trematochampsidae as a dentitionally specialized offshoot. This new taxonomic interpretation means that Africa, in any case, was involved in the biogeographical history of *Bergisuchus*. Whether the presence of trematochampsids in Africa and South America resulted from a faunal exchange across the narrow proto-Atlantic in the Upper Cretaceous or else indicates a West Gondwanan distribution prior to the separation of both continents, is open to debate. Buffetaut (1988) favoured the latter view. As for the immigration across the Tethys barrier into Europe he assumed an episode close to the Cretaceous–Tertiary boundary. The mesosuchian from El Kohol is reminiscent of trematochampsids too. It represents an endemic lineage, however, and thus would support a migrational event clearly before the Early Eocene.

The ratite bird *Palaeotis weigelti*

Palaeotis weigelti was originally described by Lambrecht (1928) on the basis of a tarsometatarsus and a pedal phalanx from the late Middle Eocene of Geiseltal, Germany, and was referred to the neognathous bustards, family Otididae. The allocation changed considerably when Houde (1986) and Houde and Haubold (1987) restudied all available specimens of *Palaeotis*, including fairly complete skeletons both from Messel and Geiseltal. They concluded that *Palaeotis* actually is a flightless, palaeognathous bird having close affinities with the ostriches. Extant birds with a particular palatal structure termed palaeognathous, comprise the orders Tinamiformes (tinamous) and Struthioniformes (ratites). Modern ratites (ostriches, rheas, cassowaries, emus, and kiwis) are flightless birds which are found on fragments of the former southern supercontinent Gondwana. The widely accepted idea of their southern origin and the subsequent isolation of these birds, with poor dispersal capabilities, by the break-up of Gondwana, is now challenged by the finding of an early, unequivocal ratite from a northern continent.

Houde and Haubold (1987) regarded *Palaeotis* as a member of the ostrich lineage on the basis of trivial derived characters; in their opinion it would make a good candidate as an actual ancestor of modern African ostriches. Consequently, *Palaeotis* was assigned to a new struthionid subfamily, Palaeotidinae, which was diagnosed by primitive features mainly of the basicranium. They advocated an exclusively Eurasian early evolution of ostriches and a subsequent dispersal to Africa sometimes during the Tertiary.

A newly discovered and virtually complete specimen from Messel (Fig. 6.1) added valuable information and modified this systematic and palaeobiogeographical view (Peters 1988*a*, *b*). Peters stressed the fact that if there were close affinities with any extant ratite family they are with the South American rheas and not with the African ostriches. He enumerated characters

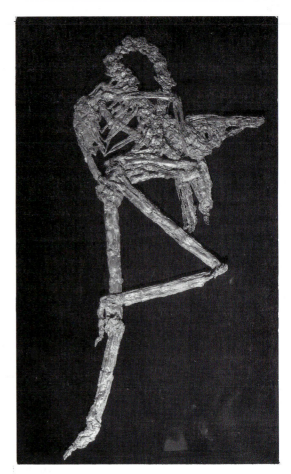

Fig. 6.1 *Palaeotis weigelti* Lambrecht, a ratite bird from the Middle Eocene oil-shales of 'Grube Messel' near Frankfurt, Germany. Various skeletal features closely resemble the extant South American rheas.

which are strongly suggestive of Rheidae: There are 7–8 pairs of ribs, only three of them being connected to the sternum; the ischium and pubis are caudally much longer than the ilium, and they are unfused; and fractures and fragments of bone on the specimen leave little doubt that the ischium and ilium were connected by a bony bar (this bridge has a different configuration in emus and cassowaries and is missing in ostriches). The scapulocoracoideum, the wing, and particularly the pelvis compare well with those of rheas among extant ratites. Since some doubt still remained about the synapomorphic character state of these features, Peters (1988*a*) raised the palaeotidines to family rank in order to house *Palaeotis*.

The ancestry of *Palaeotis* may be traced to the *Lithornis*-cohort (Houde 1986, 1988). It is a group of primitive palaeognathous birds from the Palaeocene and Eocene of North America and Europe which still had the potential of flight. There seem to be, however, no features which are demonstrably synapomorphic. On the other hand, it must be emphasized that the earliest certain ratite known to date is *Diogenornis fragilis* from the later Palaeocene of Itaborai, Brazil. It rivals the *Lithornis*-cohort biochronologically and it obviously possessed already rhea-like skeletal features (Alvarenga 1983). Thus, South America and Africa seem to be intimately involved in early ratite palaeobiogeography, and a West Gondwanan origin of *Palaeotis* (and subsequent crossing of the Tethys) appears at least as plausible as an Euramerican one. In any case, ostrich and rhea palaeobiogeography and distribution certainly cannot be explained by a simple vicariance hypothesis. Peters (1988*a*) for instance pointed out that the African Oligocene included ratites distinct from ostriches.

The phorusrhacid bird *Aenigmavis sapea* and seriemas

Phorusrhacidae is an extinct family of gruiform birds that was rather diversified in South America (Argentina, Uruguay, Brazil) during Oligocene to Pliocene times. Towards the Pliocene–Pleistocene boundary it extended its range north into Florida. The family included several giant flightless species with supposedly carnivorous habits. Phorusrhacids are closely related to the family Cariamidae and both are grouped within a higher taxon of varying rank and contents, the Cariamae. These birds represent a very old group which radiated from a primitive gruiform stock, probably prior to the break-up of Gondwanaland in Mesozoic times. No more than two cariamid species, the peculiar seriemas (*Cariama cristata*

and *Chunga burmeisteri*), survived as relics in South America.

Mourer-Chauviré (1981) reported on the first phorusrhacidan finding from Europe and its palaeobiogeographical implications. She based a new genus, *Ameghinornis*, on several bones of the shoulder-girdle and wing which were discovered from the well-known Phosphorites du Quercy, France, and date from the Late Eocene to Late Oligocene. *Ameghinornis* was a rather primitive, medium-sized bird with probably poor flight abilities. Mourer-Chauviré (1981, 1982, 1983) assumed that phorusrhacid birds evolved during the Cretaceous, either on Gondwanaland before its fragmentation or on the South American continental block shortly after its separation from Africa. The migration route to Europe and its timing are hard to judge. An early Gondwanan origin would imply the crossing of the Tethys Sea from Africa into Europe, presumably in the Late Cretaceous. If phorusrhacids originated in South America (or any other place on the southern garland including South America–Antarctica–Australia), dispersal either across Africa or else North America was involved. The African migration route could have crossed the southern proto-Atlantic via the Late Cretaceous Walvis and Rio Grande ridges, which were partially emerged then, and thereafter the Tethys barrier to Europe. The North American route could have passed the Central American isthmus in Late Cretaceous times and subsequently the northern Atlantic land corridors into Europe during the Early Eocene. Mourer-Chauviré (1982, 1983) stated that the northern route is somewhat contradicted by the absence of Palaeogene fossils of Phorusrhacidae from an otherwise fine North American record, and future discoveries from Africa will be crucial for a better understanding of Cariamae palaeobiogeography.

Peters (1987) provided valuable new information about phorusrhacidans when he described *Aenigmavis sapea* from Messel. The type specimen is virtually complete. It shows a rooster-sized bird with very strong and robust hind limbs, equipped with large, curved talons, while the wing and neck are short. Peters concluded a cursorial and carnivorous bird without, or with poor, flight capabilities. It is the geologically oldest member of the Cariamae known to date. Peters assigned *Aenigmavis* provisionally to Phorusrhacidae since he considered the subdivision of the Cariamae still unsettled. However, affinities with previously described phorusrhacids predominate (massive bones, shape of the condylus lateralis tibiotarsi and talons, short wings, lack of a pons supratendineus tibiotarsi). Besides, *Aenigmavis* shares some features with certain extant birds, the kagu (*Rhynochetos*), endemic to New Caledonia, and sunbittern (*Eurypyga*) from South and Central America (morphology of the hypotarsus, well separated condylus dorsalis ulnae), as well as the seriemas (various skull features, morphology of the olecranon). All of these birds are grouped by Cracraft (1982) within a presumably monophyletic gruiform infraorder Psophii. The results achieved from the Messel specimens to my mind strongly suggest the involvement of the Psophii (inclusive of the Cariamae) in the Mesozoic Gondwana biota and their diversification sometime in the Cretaceous.

Recently, Peters (1988*b*) also reported seriemas from Messel. The preserved hind limb and skull closely resemble the extant South American cariamids. The findings were referred to the extinct cariamid subfamily Idiornithinae, that previously was known only from the French Quercy faunas (see Mourer-Chauviré 1983). In addition to close relationships with the recent seriemas, Mourer-Chauviré (1983) stated affinities of the idiornithines with the extinct North American cariamid subfamily Bathornithinae and the peculiar extant hoatzin (*Opisthocomus*), again a bird endemic to South America! The palaeobiogeography of idiornithine birds might thus correspond with that of phorusrhacidans. In any case, the affinities between the Messel fauna and neotropical biota is emphasized by the record of cariamids.

The didelphid marsupials *Peradectes* and *Amphiperatherium*

The interrelationships and palaeobiogeography of Palaeogene marsupials are still poorly understood, and recent systematic arrangements are in disagreement on such points as generic allocations and the rank and contents of higher taxa. European didelphids compose two subfamilies (or may represent two distinct families) which experienced a rather restrained radiation. The Didelphinae includes the genera *Peratherium* and *Amphiperatherium* and the more plesiomorphic Peradectinae contains the genus *Peradectes*. Marsupials from Messel have recently become known by the finding of four virtually complete skeletons of *Peradectes* sp., *Amphiperatherium* cf. *maximum*, and an undetermined taxon (Von Koenigswald 1982; Von Koenigswald and Storch 1988). Von Koenigswald and Storch (1988) concluded that the slender *Peradectes* species with its extremely long tail was adapted arboreally while the larger short-tailed *Amphiperatherium* species was preferably a forest-floor dweller.

Peradectes and *Amphiperatherium* may have used quite different routes and time periods to reach Europe. South or North America have provided the earliest evidence of *Peradectes* chronologically (earliest Palaeocene), and Europe is third (Early Eocene). This implies a dispersal of members of this genus from North America into Europe (Crochet and Sigé 1983).

Didelphine marsupials, on the other hand, appeared penecontemporaneously in the Early Eocene on the European and North American continent, and no suitable direct ancestral stock is known from either. Thus, they must have immigrated from somewhere else and dispersed thereafter in one direction or the other. The earliest known records of European *Peratherium* and *Amphiperatherium* are from faunas such as Silveirinha and Dormaal, these are pretty close to the Palaeocene–Eocene boundary and correlate with the North American Land Mammal Age Late Clarkforkian (Godinot 1982). North American *Peratherium* (no *Amphiperatherium* is

known) appeared abruptly in the Early Wasatchian and is unknown from the preceding Clarkforkian (Krishtalka and Stucky 1983); that points
to an immigration from Europe, Asia, too, was
not a likely cradle of the ancestral stock of the
Messel *Amphiperatherium* species. The recently
discovered first Asian didelphine from the Early
Oligocene of Kasachstan (Gabunia *et al.* 1985) in
all probability was an immigrant from Europe
itself.

Obviously, southern continents were intimately
involved in early didelphine palaeobiogeography
instead. The fossil record of marsupials from
South America includes the earliest didelphines
known (latest Cretaceous or early Palaeocene of
Bolivia and Peru) and demonstrates that these
animals were abundant and taxonomically
diverse there by this time (Sigé 1972; Marshall
and de Muizon 1988). South America surely was
a major arena for early didelphine evolution and
radiations. Crochet and Sigé (1983) hypothesized
didelphine dispersal from South America into
Europe via Africa at the end of the Palaeocene, by
way of emerged intercontinental oceanic ridges
and sweepstakes paths. Yet the Early Tertiary
distribution pattern might be explained likewise
by a combination of dispersal and vicariance.
Marsupials have undergone a complex history in
Africa (Crochet 1986*a* and *b*), and an ancestral
stock of the Messel *Amphiperatherium* might
have existed before Gondwanan biota were
vicariated by continental drift during the Late
Cretaceous. The single African didelphine
known, *Peratherium africanum* from the Oligocene Jebel Qatrani Formation of Egypt, might be
evidence for that. Simons and Bown (1984), on
the other hand, strongly favoured an Eocene
European origin of *P. africanum* ancestors.

The anteater *Eurotamandua joresi*

The anteater *Eurotamandua*, family Myrmecophagidae, is one of the most spectacular Messel
fossils, both from the view of its excellent preservation and its taxonomic allocation (Storch 1981)
(Fig. 6.2). Anteaters, sloths, and armadillos are

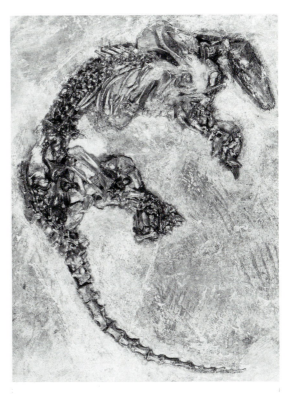

Fig. 6.2 *Eurotamandua joresi* Storch, a Middle
Eocene myrmecophagidan edentate from 'Grube Mes-
sel'. *Eurotamandua* compares well with extant South
American anteaters, in particular the genus *Tamandua*.

the living members of the formerly more diverse
order Edentata (= Xenarthra) which is highly
characteristic of the fossil and extant mammalian
faunas of South America (Hoffstetter 1982).
Edentates were thought to have flourished on the
South American continent, and deserted their
Cenozoic 'splendid isolation' (Simpson 1980) not
before the Late Miocene and in particular the
great Pan-American biotic interchange during
latest Neogene time. *Eurotamandua* does not fit
such a palaeobiogeographical scenario since it is
the only Early Tertiary anteater (suborder Vermilingua) known to date and was discovered outside the neotropical region. Previously, the fossil

history of anteaters could be traced biochronologically to the Early Miocene Santa Cruz Formation of Argentina. The single *Eurotamandua* specimen from Messel was recently complemented by isolated limb bones (humerus and ulna) from the contemporary Geiseltal Lower Coal level, Germany (Storch and Haubold 1989).

Eurotamandua was a heavily built animal, about 90 cm in total length, and perfectly adapted morphologically for the peculiar myrmecophagous feeding strategy (Storch 1981). Its preserved gut contents included insect cuticulae (presumably of termites), sand, and cemented wood of the same structure as hard carton nests of extant tree-dwelling termites (Richter 1987; Storch and Richter 1988). Among extant genera (*Myrmecophaga*, *Tamandua*, *Cyclopes*), *Eurotamandua* compares best morphologically and adaptively with the most generalized, *Tamandua*.

The skeleton exhibits synapomorphic edentate features as well as autapomorphic characters of the Vermilingua. The xenarthrous articulations of vertebrae in the posterior trunk region are unique to the edentates and are well developed in *Eurotamandua*: successive vertebrae bear supplementary articulations between a posteriorly directed anapophysis and two embracing accessory prezygapophyses on each side. The formation of a rigid pelvic assembly (= synsacrum) is another feature not known outside the edentates and likewise well developed in *Eurotamandua*: the complete synostotic fusion of the transverse processes of the posterior pseudosacral vertebra to the ischia results in a typical foramen sacroischiadicum. Other edentate features of *Eurotamandua* include the development of a secondary spine on the external scapular surface.

Within the edentates, anteater morphology can be characterized by various uniquely derived features which are shared by *Eurotamandua*. *Eurotamandua* lacks teeth. The skull is elongate and tubular, and the lower jaws are slender bony bars. The hard palate is extended posteriorly and the internal nares are displaced close to the foramen magnum. The forelimbs are highly specialized fossorially. *Eurotamandua*, like the

recent *Tamandua*, was obviously a powerful hook-and-pull digger. The third digit is enlarged. Its massive metacarpal is equipped with a characteristic dorsal muscle scar and a strong distal keel. The configuration of the distal articulating surfaces on the shortened basal phalanx is such that little or no movement was possible at the proximal interphalangeal joint. The claw of the central digit is very large and powerful. Specializations of the elbow joint unique to the extant *Tamandua* and *Myrmecophaga* (Taylor 1978, 1985) are likewise found in *Eurotamandua* (Storch and Haubold 1989). The entepicondyle of the humerus is markedly extended and exhibits a long straight medial margin and a very wide and rather deep distal notch for the passage of forearm muscles, which is in line with the axis of rotation of the elbow joint. On the ulna, a bill-shaped anconaeal process above the semilunar notch is absent and the radial facet is not expanded into a tongue-shaped lateral process. Within the Vermilingua, *Eurotamandua* even had a very distinctive characteristic of the Myrmecophagidae (*Myrmecophaga* and *Tamandua*) which separates them from the Cyclothuridae (*Cyclopes*)—a supplementary bulla tympanica in front of the tympanic cavity (Storch and Habersetzer 1991).

Extant anteaters show little diversity, and the myrmecophagidan morphology and adaptive type can be traced to 50 my old *Eurotamandua*. The earliest record of undisputed edentates date from the later Palaeocene of South America, and these remains are clearly referable to armadillos, probably of the dasypodan and glyptodontan branches within that group (Cifelli 1983; Engelmann 1985). Intermediates among the three easily recognized adaptive types of xenarthrans—anteaters, armadillos, and sloths—are unknown from fossils. All this suggests that the phylogenetic separation of the major edentate lineages considerably predates their first occurrence in the fossil record, and that the origin of xenarthrans extends well back into the Cretaceous. This view is corroborated by albumin immunological evidence, indicating that the xenarthran families separated at least 75–80 my ago (Sarich 1985).

Extant edentates have retained a suite of primitive features which suggest that they are the most primitive eutherian order and a sister group to all other eutherians (McKenna 1975; Engelmann 1985). The general primitiveness and hence the remote divergence of edentates from other eutherians are provided by a combination of physiological, molecular, and anatomical characteristics. The latter include the pattern of carotid circulation and the female reproductive tract. Here again, the view of a very early origin of edentates during the Cretaceous, supposedly not too far from the basal marsupial–eutherian split, is corroborated.

South America was a major theatre for Tertiary edentate radiations and it is at least probable that edentates originated on Gondwanaland. The last direct connection of South America to Africa was broken probably not before the later Cretaceous about 90 my ago (Reyment and Dingle 1987) and true oceanic conditions throughout the South Atlantic between South America and Africa were installed probably not before the latest Cretaceous (Sclater *et al.* 1977). New geological, geophysical, and invertebrate palaeontological evidence suggest that the final opening of the South Atlantic was more complex than previously thought and intermittent land connections between Brazil and West Africa most probably existed through the Maastrichtian or even the Early Palaeocene (Stinnesbeck 1991). From the above evidence I concluded (Storch 1984, 1986) that edentates were in existence and diversified before the final break-up of the South American and African land masses and that Africa was then part of the myrmecophagidan distributional area. Accordingly, the presence of *Eurotamandua* in central Europe is attributed to a transtethyan crossing from Africa. Admittedly, no myrmcophagidan fossils are known from Africa. We must bear in mind, however, that the Late Cretaceous/ Early Palaeogene fossil record from Africa is still rather blank, and that, until quite recently, African marsupials were unknown.

CONCLUSIONS

The Middle Eocene vertebrate fauna from Messel has a distinctive South American appearance shown by the occurrence of a rhealike ratite, phorusrhacid and cariamid birds, and a myrmecophagidan edentate. Didelphine marsupials and a trematochampsid mesosuchian, too, seem to have originated from an indigenous Gondwana stock. The Early Tertiary distribution patterns can be interpreted by vicariance, dispersal, or a combination of the two. One hypothesis is that these groups originated in South America as a result of vicariance caused by the break-up of Gondwana and then dispersed by direct overland passage and/or sweepstakes routes into Europe via Africa. Recently Stinnesbeck (1991) stressed the point that transgression–regression events with island chains and even landbridges occurred between Brazil and West Africa until the Maastrichtian, and interchanges of terrestrial vertebrates may have occurred until the Early Palaeocene. Further south, the Rio Grande Rise and Walvis Ridge provided a sizeable island chain between the separating continents South America and Africa in Late Cretaceous through Palaeocene time (Reyment 1980). A limited faunal exchange between Africa and Europe across the Tethys barrier obviously occurred around the Cretaceous–Tertiary boundary (Gheerbrant 1987) and, intermittently, during the Palaeogene (Mahboubi *et al.* 1986). Didelphinae and Cariamae may have migrated along this route. A second hypothesis favours a dispersal route from South America via North America into Eurasia; it is somewhat contradicted by the absence of relevant fossils from a rich North American record. Recently, Gingerich (1985) supported the idea of an edentate dispersal from South America across North America into Asia by late Palaeocene time. His evidence, however, is essentially based on North American Palaeanodonta, an order of disputed xenarthran relationships.

I seriously consider, however, a somewhat different palaeogeographical view, and I strongly advocate that view to explain the occurrence of the Messel anteater (Storch 1984, 1986). Students of the above vertebrates are in full agreement that these animals represent very primitive groups and that their origin can be traced, most probably, to the Cretaceous. The diversification of some of these vertebrates could have preceded the final break-up of Gondwana, so that the ancestral stocks of the Messel species could have reached Europe from Africa as part of an original distribution area. An improvement in the fossil record of the Late Cretaceous–Early Palaeocene of Africa will be crucial for a better understanding of African–South American faunal affinities and its effects upon Early Tertiary faunal development in the Holarctic.

Finally, there is no evidence from fossils, thus far, that primitive early groups, such as anteaters, once may have had a worldwide distribution and did survive only in South America's Tertiary 'splendid isolation'. Likewise, a holarctic origin of these typical 'Southerners' followed by a dispersal opposite to that considered above, cannot be totally dismissed but is less reasonable.

REFERENCES

Alvarenga, H. M. F. (1983). Uma ave ratitae do Paleoceno Brasileiro: bacia calcária de Itaboraí, Estado do Rio de Janeiro, Brazil. *Boletim do Museu Nacional Rio de Janeiro, N. S., Geologia*, No. 41, 1–7.

Berg, D. E. (1966). Die Krokodile, insbesondere *Asiatosuchus* und aff. *Sebecus* ? aus dem Eozän von Messel bei Darmstadt/Hessen. *Abhandlungen des Hessischen Landesamtes für Bodenforschung*, Wiesbaden, **52**, 1–105.

Buffetaut, E. (1980). Histoire biogéographique des Sebecosuchia (Crocodylia, Mesosuchia): un essai d'interprétation. *Annales de Paléontologie (Vertébrés)*, **66**, 1–18.

Buffetaut, E. (1982). A ziphodont mesosuchian crocodile from the Eocene of Algeria and its implications for vertebrate dispersal. *Nature*, **300**, 176–8.

Buffetaut, E. (1988). The ziphodont mesosuchian crocodile from Messel: a reassessment. *Courier Forschungsinstitut Senckenberg*, Frankfurt a.M., **107**, 211–21.

Cifelli, R. L. (1983). Eutherian tarsals from the late Paleocene of Brazil. *American Museum Novitates*, **2761**, 1–31.

Cracraft, J. (1982). Phylogenetic relationships and transantarctic biogeography of some gruiform birds. In *Phylogénie et paléobiogéographie. Livre jubilaire en l'honneur de Robert Hofstetter* (ed. E. Buffetaut, P. Janvier, J. C. Rage, and P. Tassy). *Geobios*, Lyon, Mémoire spécial No. 6, pp. 393-402.

Crochet, J.-Y. (1986*a*). *Kasserinotherium tunisiense* nov. gen., nov. sp., troisième marsupial découvert en Afrique (Eocène inférieur de Tunisie). *Comptes rendus de l'Académie des Sciences de Paris*, II, **302**, 923–6.

Crochet, J.-Y. (1986*b*). Le berceau des marsupiaux. *Le Recherche*, **17**, No. 174, 275–6.

Crochet, J.-Y. and Sigé, B. (1983). Les mammifères Montiens de Hainin (Paléocène moyen de Belgique) Part III: marsupiaux. *Palaeovertebrata*, Montpellier, **13**(3), 51–64.

Engelmann, G. F. (1985). The phylogeny of the Xenarthra. In *The evolution and ecology of armadillos, sloths, and vermilinguas* (ed. G. G. Montgomery), pp. 51–64. Smithsonian Institution Press, Washington.

Franzen, J. L. (1990). Grube Messel. In *Palaeobiology. A synthesis* (ed. D. E. G. Briggs and P. R. Crowther), pp. 289–94. Blackwell Scientific Publications, Oxford.

Franzen, J. L. and Haubold, H. (1986). The middle Eocene of European mammalian stratigraphy. Definition of the Geiseltalian. *Modern Geology*, **10**, 159–70.

Gabunia, L. K., Shevyreva, N. C., and Gabunia, U. D. (1985). Über den ersten Fund eines Marsupialiers in Asien. *Moskau Akademii Nauk SSSR*, **281**, 684–5.

Gheerbrant, E. (1987). Les vertébrés continentaux de l'Adrar Mgorn (Maroc, Paléocène); une dispersion de mammifères transtéthysienne aux environs de la limite mésozoïque/cénozoïque? *Geodinamica Acta*, Paris, **1**(4/5), 233–46.

Gingerich, P. D. (1985). South American mammals in the Paleocene of North America. In *The great American biotic interchange* (ed. F. G. Stehli and S. D. Webb), pp. 123–37. Plenum Publishing Corporation, New York.

Godinot, M. (1982). Aspects nouveaux des échanges entre les faunes mammaliennes d'Europe et d'Amérique du Nord a la base de l'Éocène. In *Phylogénie et paléobiogéographie. Livre jubilaire en l'honneur de Robert Hofstetter* (ed. E. Buffetaut, P. Janvier, J. C.

Rage, and P. Tassy). *Geobios*, Lyon, Mémoire spécial No. 6, pp. 404–12.

Hofstetter, R. (1982). Ls edentes xenarthres, un groupe singulier de la faune neotropicale (originne, affinités, radiation adaptative, migrations et extinctions). In *Proceedings of the First International Meeting on Palaeontology, Essentials of Historical Geology, Venice 2–4 June 1981* (ed. E. M. Gallitelli), pp. 385–443. S.T.E.M. Mucchi, Modena.

Houde, P. (1986). Ostrich ancestors found in the Northern Hemisphere suggest new hypothesis of ratite origins. *Nature*, **324**, 563-5.

Houde, P. (1988). Paleognathous birds from the early Tertiary of the northern hemisphere. *Publications of the Nuttall Ornithological Club*, No. 22, vii and 1–148. Cambridge, Massachusetts.

Houde, P. and Haubold, H. (1987). *Palaeotis weigelti* restudied: a small middle Eocene ostrich (Aves: Struthioniformes). *Palaeovertebrata*, Montpellier, **17**(2), 27–42.

Koenigswald, W. von (1982). Die erste Beutelratte aus dem mitteleozänen Ölschiefer von Messel bei Darmstadt. *Natur und Museum*, Frankfurt a.M., **112**(2), 41–8.

Koenigswald, W. von and Storch, G. (1988). Messeler Beuteltiere–unauffällige Beutelratten. In *Messel— Eine Schaufenster in die Geschichte der Erde und des Lebens* (ed. S. Schaal and W. Ziegler), pp. 153–8. Waldemar Kramer, Frankfurt a.M.

Krishtalka, L. and Stucky, R. K. (1983). Paleocene and Eocene marsupials of North America. *Annals of Carnegie Museum*, **52**(10), 229-63.

Kuhn, O. (1968). *Die vorzeitlichen Krokodile*. Krailling bei München.

Lambrecht, K. (1928). *Palaeotis weigelti* n. g. n. sp., eine fossile Trappe aus der mitteleozänen Braunkohle des Geiseltales. *Jahrbuch des Halleschen Verbandes für die Erforschung der mitteldeutschen Bodenschätze und ihre Verwertung, N.F.*, **7**, 20–9.

McKenna, M. C. (1975). Toward a phylogenetic classification of the Mammalia. In *Phylogeny of the primates. A multidisciplinary approach* (ed. W. P. Luckett and F. S. Szalay), pp. 21–46. Plenum Press, New York.

Mahboubi, M., Ameur, R., Crochet, J. Y., and Jaeger, J. J. (1986). El Kohol (Saharan Atlas, Algeria): a new Eocene mammal locality in northwestern Africa. Stratigraphical, phylogenetic and paleobiogeographical data. *Palaeontographica*, Stuttgart, Abt. A, **192**(1–3), 15–49.

Marshall, L. G. and de Muizon, C. (1988). The dawn of the age of mammals in South America. *National Geographic Research*, **4**(1), 23-55.

Mourer-Chauviré, C. (1981). Première indication de la présence de Phorusracidés, famille d'oiseaux géants d'Amérique du Sud, dans le Tertiaire Européen: *Ameghinornis* nov. gen. (Aves, Ralliformes) des Phosphorites du Quercy, France. *Géobios*, Lyon, **14**(5), 637–47.

Mourer-Chauviré, C. (1982). Les oiseaux des Phosphorites du Quercy (Eocène supérieur à Oligocène supérieur): implications paléogéographiques. In *Phylogénie et paléobiogéographie. Livre jubilaire en l'honneur de Robert Hofstetter* (ed. E. Buffetaut, P. Janvier, J. C. Rage, and P. Tassy). *Geobios*, Lyon, Mémoire spécial No. 6, pp. 413–26.

Mourer-Chauviré, C. (1983). Les Gruiformes (Aves) des Phosphorites du Quercy (France) 1. sous-ordre Cariamae (Cariamidae et Phorusrhacidae) systématique et biostratigraphie. *Palaeovertebrata*, Montpellier, **13**(4), 83–143.

Peters, D. S. (1987). Ein 'Phorusrhacide' aus dem Mittel-Eozän von Messel (Aves: Gruiformes: Cariamae). In *L'Evolution des oiseaux d'après le témoignage des fossiles, Table ronde internationale du CNRS, Lyon-Villeurbanne 18–21 Septembre 1985* (ed. C. Mourer-Chauviré), pp. 71–87. Documents des Laboratoires de Géologie Lyon, No. 99.

Peters, D. S. (1988a). Ein vollständiges Exemplar von *Palaeotis weigelti* (Aves, Palaeognathae). *Courier Forschungsinstitut Senckenberg*, Frankfurt a.M., **107**, 223–33.

Peters, D. S. (1988b). Die Messel-Vögel—eine Landvogelfauna. In *Messel—ein Schaufenster in die Geschichte der Erde und des Lebens* (ed. S. Schaal and W. Ziegler), pp. 134–51. Waldemar Kramer, Frankfurt a.M.

Rage, J.-C. (1981). Les continents Péri-atlantiques au Crétacé supérieur: migrations des faunes continentales et problèmes paléogéographiques. *Cretaceous Research*, **2**, 65–84.

Reyment, R. A. (1980). Paleo-oceanology and paleobiogeography of the Cretaceous South Atlantic Ocean. *Oceanologica Acta*, **3**(1), 127–33.

Reyment, R. A. and Dingle, R. V. (1987). Palaeogeography of Africa during the Cretaceous period. *Palaeogeography, Palaeoclimatology, Palaeoecology*, **59**, 93–116.

Richter, G. (1987). Untersuchungen zur Ernährung eozäner Säuger aus der Fossilfundstätte Messel bei

Darmstadt. *Courier Forschungsinstitut Senckenberg*, Frankfurt a.M., **91**, 1–33.

Sarich, V. M. (1985). Xenarthran systematics: albumin immunological evidence. In *The evolution and ecology of armadillos, sloths, and vermilinguas* (ed. G. G. Montgomery), pp. 77–81. Smithsonian Institution Press, Washington.

Schaal, S. and Ziegler, W. (ed.) (1988). *Messel—ein Schaufenster in die Geschichte der Erde und des Lebens*. Waldemar Kramer, Frankfurt a.M.

Schmidt-Kittler, N. (ed.) (1987). International symposium on mammalian biostratigraphy and paleoecology of the European Paleogene—Mainz, February 18th–21st 1987. *Münchner geowissenschaftliche Abhandlungen, Geologie und Paläontologie*, A10, pp. 1–312. Pfeil, München.

Sclater, J. G., Hellinger, S., and Tapscott, C. (1977). The palaeobathymetry of the Atlantic Ocean from the Jurassic to the present. *The Journal of Geology*, **85**, 509–52.

Sigé, B. (1972). La faunule de mammifères du Crétacé supérieur de Laguna Umayo (Andes péruviennes). *Bulletin du Muséum national d'Histoire Naturelle, Sciences de la Terre*, **19**, 3ième serie, No. 99, 375–405.

Simons, E. L. and Bown, T. M. (1984). A new species of *Peratherium* (Didelphidae; Polyprotodonta): the first African marsupial. *Journal of Mammalogy*, **65**(4), 539–48.

Simpson, G. G. (1980). *Splendid isolation. The curious history of South American mammals*. Yale University Press, New Haven.

Stinnesbeck, W. (1991). Did the South American elements of the Messel fauna migrate via Africa? *Abstracts International Conference Monument Grube Messel*, Hessisches Landesmuseum, Darmstadt.

Storch, G. (1981). *Eurotamandua joresi*, ein Myrmecophagide aus dem Eozän der 'Grube Messel' bei Darmstadt (Mammalia, Xenarthra). *Senckenbergiana lethaea*, Frankfurt a.M., **61**, 247–89.

Storch, G. (1984). Die alttertiäre Säugetierfauna von Messel–ein paläobiogeographisches Puzzle. *Naturwissenschaften*, **71**, 227–33.

Storch, G. (1986). Die Säuger von Messel: Wurzeln auf vielen Kontinenten. *Spektrum der Wissenschaft*, 1986, (6), 48–65.

Storch, G. and Habersetzer, J. (1991). Rückverlagerte Choanen und akzessorische Bulla tympanica bei rezenten Vermilingua und *Eurotamandua* aus dem Eozän von Messel. *Zeitschrift für Säugetierkunde*, **56**, 257–71.

Storch, G. and Haubold, H. (1989). Additions to the Geiseltal mammalian faunas, middle Eocene: Didelphidae, Nyctitheriidae, Myrmecophagidae. *Palaeovertebrata*, Montpellier, **19**(3), 95–114.

Storch, G. and Richter, G. (1988). Der Ameisenbär *Eurotamandua*—ein 'Südamerikaner in Europa. In *Messel—ein Schaufenster in die Geschichte der Erde und des Lebens* (ed. S. Schaal and W. Ziegler), pp. 209–15. Waldemar Kramer, Frankfurt a.M.

Storch, G. and Schaarschmidt, F. (1988). Fauna and Flora von Messel–ein biogeographisches Puzzle. In *Messel—ein Schaufenster in die Geschichte der Erde und des Lebens* (ed. S. Schaal and W. Ziegler), pp. 291–7. Waldemar Kramer, Frankfurt a.M.

Taylor, B. K. (1978). The anatomy of the forelimb in the anteater (*Tamandua*) and its functional implications. *Journal of Morphology*, **157**, 347–68.

Taylor, B. K. (1985). Functional anatomy of the forelimb in vermilinguas (anteaters). In *The evolution and ecology of armadillos, sloths, and vermilinguas* (ed. G. G. Montgomery), pp. 163–71. Smithsonian Institution Press, Washington.

7 FOSSIL AMPHIBIANS AND REPTILES AND THE AFRICA–SOUTH AMERICA CONNECTION

Eric Buffetaut and Jean-Claude Rage

INTRODUCTION

Although it was the subject of much controversy among geologists and palaeontologists until well into the 1960s, the former existence of a land connection between Africa and South America no longer needs to be demonstrated. Within the general framework of plate tectonics and continental drift, it is now clear that Africa and South America were once part of the southern supercontinent Gondwana, itself a constituent of Pangaea during the Late Palaeozoic and the Early Mesozoic. The distribution of some fossil amphibians and reptiles was one of the important palaeobiogeographical arguments in favour of such reconstructions. The occurrence of very closely related representatives of a group of small early reptiles, the mesosaurs, in the Permian of South America and South Africa, for instance, was used as evidence of a connection between Africa and South America in the Late Palaeozoic, well before continental drift became generally accepted. The Late Triassic dinosaur faunas of South Africa and South America also exhibit resemblances—which is not surprising in a palaeobiogeographical situation of global faunal similarity, due to the existence of Pangaea. Although it may seem paradoxical, the Africa–South America connection becomes an interesting scientific problem mainly when it ceases to exist, when the opening of the proto-Atlantic Ocean sometime in the Cretaceous separates the two continents from each other and makes faunal interchange increasingly difficult. A comparative history of amphibians and reptiles in South America and Africa, through the Cretaceous and into the Tertiary, reveals similarities and divergences suggesting a pattern of faunal exchanges in which vicariant evolution and dispersal across barriers both played a part.

BEFORE THE SPLIT: EARLY CRETACEOUS FAUNAS

The study of non-marine fossil vertebrates has provided a significant contribution to the dating of the separation of Africa from South America by the opening of the proto-Atlantic Ocean. The occurrence of very similar faunal elements among terrestrial vertebrate assemblages from the Lower Cretaceous of Africa and South America suggests that the faunas of those continents had not yet been isolated from each other by a marine barrier. Dating the latest 'common' African–South American non-marine vertebrates fauna thus constrains the date of the establishment of a continuous seaway between Africa and South America.

The Early Cretaceous mammals of Africa and South America still largely unknown, the most useful group of tetrapods, from a palaeobiogeographical point of view, has proved to be the reptiles. Evidence strongly suggesting faunal continuity among African and South American reptiles in the Early Cretaceous has been provided by a comparison between the fauna from the Gadoufaoua area in Niger and those from various localities in Brazil.

Crocodylia

Two genera of mesosuchian crocodilians are currently known to have been present in both South

America and Africa in the Early Cretaceous (Buffetaut 1985*a*).

Sarcosuchus de Broin and Taquet 1966 is a very large (up to about 11 m long) freshwater representative of the family Pholidosauridae. The type species of the genus, *Sarcosuchus imperator*, was described by de Broin and Taquet (1966) from the Elrhaz Formation at Gadoufaoua, in Niger, which is usually referred to the Aptian (Taquet 1976); *S. imperator* is also known to occur in deposits of roughly the same age in Algeria and Tunisia (de Broin and Taquet 1966). It was later shown (Buffetaut and Taquet 1976, 1977*a,b*) that remains of a very large crocodilian reported as early as 1860 by Allport, and later by Mawson and Woodward (1907), from the Bahia basin of Brazil in fact belonged to the genus *Sarcosuchus*, although first called *Crocodilus hartti* by Marsh (1869) and then referred to the genus *Goniopholis* by Mawson and Woodward (1907). Whereas the African form is represented by fairly complete skulls and postcranial elements from Gadoufaoua, the Brazilian form is represented by more fragmentary specimens, including part of a mandibular symphysis; this makes comparisons difficult, but the differences between the two forms appear to be slight and may not justify a separation at the species level (in which case the name *Sarcosuchus hartti* should prevail). The Brazilian specimens apparently were collected from the Ilhas Formation, a part of the Bahia Series which may be slightly older than the Aptian; there may thus be a slight age difference between the known African and South American occurrences of *Sarcosuchus*.

A second non-marine crocodilian genus common to Africa and South America in the Early Cretaceous is *Araripesuchus* Price 1959. It was first described on the basis of a skull from the Santana Formation of north-eastern Brazil by Price (1959), with *A. gomesii* as type-species. The vertebrate-bearing part of the Santana Formation, which is well known for its remains of fish and pterosaurs preserved in nodules, is referred to the Upper Aptian (Brito and Campos 1983). *Araripesuchus* belongs to a family of small terres-

trial mesosuchians, the Uruguaysuchidae, otherwise known from other Cretaceous localities in South America. In 1979, the occurrence of *Araripesuchus* in the Elrhaz Formation of Gadoufaoua was reported by Buffetaut and Taquet. The African form, represented by a partial skull and various postcranial elements, differs from the type species by a few details and has been considered as a separate species, *Araripesuchus wegeneri*, by Buffetaut (1981), but there is no doubt that the two forms are very closely related. The markedly terrestrial adaptations of *Araripesuchus* make it especially unlikely that this small crocodilian could cross marine barriers.

Chelonia

Close resemblances between Early Cretaceous turtles from Brazil and Niger have been reported by de Broin (1980, 1988). The pleurodiran *Araripemys*, first described from the Santana Formation of north-eastern Brazil by Price (1973), with the type species *A. barretoi*, is represented at Gadoufaoua by specimens referred to as *Araripemys* sp. (de Broin 1980). In addition, remains from the Santana Formation of Brazil are referred to as aff. *Teneremys* sp. by de Broin (1988), *Teneremys* being a pelomedusid turtle first described by de Broin (1980) from Gadoufaoua, with *T. lapparenti* as type species. According to de Broin (1988), all known Early Cretaceous pleurodirans are freshwater forms from Brazil and Africa.

Other reptiles

Taken together, the evidence based on crocodilians and chelonians from the Aptian Elrhaz Formation of Niger and several formations of roughly the same age in Brazil (notably the Santana Formation) indicates that, at least among those groups, there were very close faunal similarities between those regions. It is, therefore, unlikely that they could have been separated by a continuous marine barrier. It should also be emphasized that the evidence based on crocodilians and turtles is supported by that based on other groups of organisms, including ostracodes

(Krömmelbein 1971) and coelacanth fish (Wenz 1980). Moreover, several groups of vertebrates are well represented in the Lower Cretaceous of either Africa or South America, but still poorly known on the other side of the Atlantic, so that no meaningful comparisons can currently be made. Dinosaurs are a good example of such a group: whereas abundant and sometimes well-preserved dinosaur specimens are known from various Early Cretaceous formations in Africa (de Lapparent 1960; Taquet 1976), very little is known of the contemporaneous dinosaurs of South America (although a 'dicraeosaurid' sauropod has recently been reported from the Neocomian of Patagonia by Bonaparte (1986); it is said to be clearly reminiscent of *Dicraeosaurus* from the Upper Jurassic of East Africa.) A possible 'dicraeosaurine' has been reported from Gadoufaoua by Taquet (1976). Mawson and Woodward (1907) reported the occurrence of vertebral centra which 'seem to agree closely with the corresponding bones of Iguanodonts' from the Bahia basin. Several iguanodontids, represented by well-preserved skeletons, are known from the Aptian of Gadoufaoua (Taquet 1976), but it is doubtful whether the vertebral centra from Brazil would be sufficient to reveal significant resemblances. Nevertheless, extrapolation from what is known of crocodilians and turtles allows fairly safe predictions about which hitherto unrecorded groups of dinosaurs should be expected to occur in Aptian and older deposits in South America; they include iguanodontids and peculiar theropods of the family Spinosauridae, both families being recorded from the Aptian of Niger. Finally, as will be shown below, some vertebrate groups which are known to occur in both Africa and South America in the Upper Cretaceous were probably present on the united African–South American land mass before the opening of the proto-Atlantic; the pipid frogs and the madtsoiid snakes are among them, as well as the trematochampsid crocodilians, which are known to occur in the Early Cretaceous of South America but have not yet been recorded from deposits older than the Late Cretaceous in Africa.

On the basis of the evidence listed above, it is reasonable to assume that in the Aptian the South American and African non-marine vertebrate faunas had not yet been separated by a continuous marine barrier. This is in good agreement with evidence based on the distribution of marine invertebrates, especially ammonites, which indicates that a connection between the proto-South Atlantic and the proto-central Atlantic did not become established until the Late Albian (Förster 1978). However, the newly-formed South Atlantic was at first a narrow seaway, and continuing resemblances between the amphibian and reptile faunas of Africa and South America after its establishment, in the Late Cretaceous and even later, poses some interesting problems: are such resemblances the result of continuing interchanges across what may have been, at least at

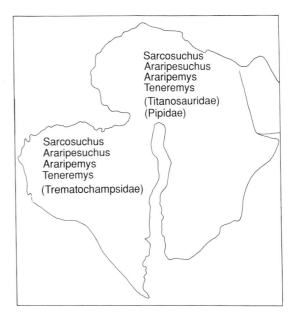

Fig. 7.1 Resemblances between the amphibian and reptile faunas of Africa and South America in the Aptian, before the establishment of a continuous seaway between the two continents. Genera of crocodilians and chelonians common to both areas are listed. The names in parentheses are those of families known from fossil evidence on one continent only, but which were very probably present on both.

times, a relatively inefficient barrier, or can they be explained as a consequence of a common Early Cretaceous origin in slowly evolving groups?

AFTER THE SEPARATION: LATE CRETACEOUS AND CAINOZOIC FAUNAS

Evidence for faunal connections between Africa and South America in the Late Cretaceous, either through interchange or as a result of vicariant evolution as discussed above, is based on groups of anuran amphibians, snakes, crocodilians, and dinosaurs.

Amphibia (Pipidae)

Pipidae represent an early offshoot of anurans. They are aquatic freshwater frogs living today in Africa (south of the Sahara) and South America (north-eastern tropical lowlands; one species reaches Panama). Fossils are known only from Africa, or more accurately from the African plate, and South America.

The Pipidae comprise five living genera (with more than 30 species): *Xenopus*, *Silurana*, *Hymenochirus*, and *Pseudhymenochirus* in Africa, and *Pipa* in South America. The *Silurana* species have been included in *Xenopus* for a long time, but Cannatella and Trueb (1988*a*) have separated these two genera. *Protopipa* and *Hemipipa*, which are sometimes considered as distinct genera, are now synonymized with *Pipa*. *Xenopus* and *Silurana* form two independent phyletic lines (Xenopodinae and Siluraninae) today restricted to Africa, whereas *Hymenochirus*, *Pseudhymenochirus*, and *Pipa* make up a monophyletic assemblage (Pipinae) present in both Africa and South America (Cannatella and Trueb 1988*b*). According to Cannatella and Trueb, the xenopodine branch diverged first, followed by the Siluraninae. It is of interest to note that they consider the Neotropical-African Pipinae as the more advanced group within the family. The pipid interrelationships put forward by Canna-

tella and Trueb (1988*a*) appear to be the best established. Bisbee *et al.* (1977) have proposed a phyletic tree based on albumin data; they indicated that the family originated at least 130 my ago, which seems quite probable; unfortunately, they did not estimate the age of the divergence between the Neotropical *Pipa* and African *Hymenochirus* lineages.

The oldest known pipid fossils come from the Early Cretaceous of Israel, that is, a part of the African plate: *Thoraciliacus* and *Cordicephalus* from the Hatira Sandstone at Makhtesh Ramon (Nevo 1968), and *Shomronella*, represented only by tadpoles (Estes *et al.* 1978), from the Tayasir Volcanics at Wadi el Malih; both formations are of Hauterivian–Barremian age, the Hatira Sandstone being younger than the Tayasir Volcanics (Mimran 1972; Estes *et al.* 1978). In Africa, the Late Cretaceous (Early Senonian) of Niger has yielded two pipid genera: one of them is a pipine frog, whereas the second form is *Xenopus*-like (Baez and Rage 1988; Baez and Rage in preparation). *Eoxenopoides*, from the Late Eocene or Oligocene of South Africa appears as a remote relative of *Xenopus* (Estes 1977). Ahl (1926) described *Xenopus stromeri* from the Miocene of Namibia. *Xenopus* sp. (or perhaps *Silurana* sp.) was reported from the Middle Miocene of Morocco (Vergnaud-Grazzini 1966), and the Late Miocene of Oued Zra (Morocco) has yielded an unpublished *Xenopus* (or *Silurana*) sp.

In South America, the oldest pipid is *Saltenia* from the Santonian–Campanian of Alemanía, north-eastern Argentina. According to Baez (1981), the phyletic line leading to *Saltenia* originates after the *Cordicephalus–Thoraciliacus* one but before all other lineages; in other words, *Saltenia* represents a very primitive pipid phyletic branch. A fossil species from the Middle Palaeocene of Itaboraí, south-eastern Brazil, has been referred to the living genus *Xenopus* (*X. romeri*; Estes 1975); according to Estes, *X. romeri* is closely related to the living *X. tropicalis*, which means that this Palaeocene species should now be referred to the living genus *Silurana*. *Shelania pascuali* comes from the Late Palaeocene–Early

Eocene of Laguna del Hunco (southern Argentina); Estes (1975) referred *Shelania* to the synonymy of *Xenopus* (i.e. at that time, *Xenopus* + *Silurana*), but Estes and Baez (1985) have retained the extinct *Shelania* as a distinct genus. Anyhow, *Shelania* seems to be close to the primitive African pipid lineages. Unpublished Pipidae from the Palaeocene–Eocene of South America were reported by Estes and Baez (1985).

The stratigraphical and geographical distribution of fossils does not definitely indicate whether the family spread over the still connected Africa–South America block, or crossed the incipient South Atlantic Ocean. The oldest known African fossils slightly antedate the opening of the South Atlantic, but the oldest known South American pipid is clearly posterior to this event. However, *Saltenia*, the oldest South American pipid, quite probably represents a phyletic line which diverged very early from the main pipid line of evolution (Baez 1981), before the *Xenopus* line arose. Estes (1975) supposed that the diversification of *Xenopus* (i.e. *Xenopus* + *Silurana*) should have antedated the separation of Africa and South America, which seems probable; if so, the early pipids spread over the still united Africa–South America block, but available palaeontological data do not provide evidence which would corroborate this inference. Therefore, a late dispersal across the incipient South Atlantic, although improbable, cannot be absolutely ruled out.

On the other hand, the close relationships that link *Silurana romeri* from the Brazilian Palaeocene and the living African *Silurana tropicalis* argue for a rather late dispersal of members of the Siluraninae from Africa to South America, that is, a dispersal across the Late Cretaceous (and/or Palaeocene) South Atlantic. This led Estes (1975) to suppose that *S. romeri* was rafted to South America across the still narrow ocean in spite of the salt intolerance of anurans. He imagined peculiar rafts in which the salt concentration would have been sufficiently low; such rafts would have permitted a crossing of the South Atlantic barrier during the Late Cretaceous, that is, before

the oceanic gap had increased too much. But it is now known that during the Late Cretaceous, and perhaps the Palaeocene, the Walvis and Rio Grande Rises, located between Africa and South America, were at least partly emerged and probably acted as a filter bridge (Rage, 1988). According to Reyment (1980), various organisms crossed the South Atlantic via this route. Anurans could also have used it, which would have permitted them to escape the insuperable problems caused by marine water. *Shelania* probably provides another evidence of such a late crossing.

Therefore, two phases may be distinguished: (1) an early dispersal, probably prior to the split between Africa and South America (or perhaps slightly after this event, across the incipient South Atlantic); (2) a late dispersal, after the separation of the two continents, by Late Cretaceous (and perhaps Palaeocene) times, which allowed African evolutionary lines to reach South America (*Silurana romeri* and probably *Shelania*); these pipids of African origin were unable to colonize South America durably.

Reptilia

Serpentes

Madtsoiidae:
Madtsoiid snakes form a primitive group of small to gigantic forms defined by the presence of parazygantral foramina on the posterior border of the neural arch, a very characteristic derived character. The Madtsoiidae have long been considered as typically Gondwanan, but typical madtsoiid vertebrae have recently been found in the Maastrichian of Laño (Spanish Basque Country). The madtsoiids are known from the Latest Cretaceous (Campanian–Maastrichtian)—Early Eocene in South America, the Late Cretaceous (Early Senonian)—Late Eocene in Africa, the Late Cretaceous (Santonian–Campanian) in Madagascar, and the Pleistocene in Australia (Rage 1984, 1987; Albino 1986).

The oldest known snakes do not antedate the Albian (Cuny *et al.* 1990); on the other hand, the oldest known Madtsoiidae may be Coniacian at the earliest (Early Senonian of In Beceten, Niger; de Broin *et al.* 1974). It is not possible to estimate the date of origin of this family. Obviously, snakes originated before the Albian, that is, prior to the separation of Africa and South America, but the Madtsoiidae may have arisen before or after the opening of the South Atlantic; in other words, it is not possible to establish whether the presence of this family in both Africa and South America results from a vicariant event or from dispersal across the oceanic gap.

Apart from this early history, stratigraphical and geographical distribution suggests a later dispersal between Africa and South America. The family comprises two distinct assemblages:

1. large and morphologically very homogeneous forms, represented by *Madtsoia* and *Gigantophis*
2. small forms, *Alamitophis*, *Patagoniophis*, and *Rionegrophis*, the latter being only doubtfully referred to the family.

In Africa, only large Madtsoiidae are known: *Madtsoia* aff. *M. madagascariensis* from the Early Senonian of In Beceten (southwestern Niger) and *Gigantophis garstini* from the Late Eocene of 'the Fayûm' (northern Egypt). An indeterminate madtsoiid (vertebrae of juveniles only) comes from the Late Palaeocene of Morocco (Rage 1987; Gheerbrant 1987). In South America, unequivocal Late Cretaceous Madtsoiidae are all small snakes from the Campanian–Maastrichtian of Cerro Cuadrado, southern Argentina (Albino 1986): *Alamitophis*, *Patagoniophis*, and *Rionegrophis* if the latter is a madtsoiid. The Earliest Cainozoic of Tiupampa (Bolivia) has yielded the centrum of a large vertebra which could belong either to a boid or to a madtsoiid snake (Rage, 1991). If this vertebra actually belongs to a madtsoiid, then it could be referred to the *Madtsoia*–*Gigantophis* group and it would be the oldest South American representative of this assemblage. Cainozoic South American Madtsoiidae all belong to the genus *Madtsoia*: *Madtsoia* sp. from the Middle Palaeocene of Itboraí (southeastern Brazil), *Madtsoia* cf. *M. bai* from the Late Palaeocene of

Cerro Pan de Azucar (southern Argentina), and *Madtsoia bai* from the Early Eocene of Cañadón Vaca (southern Argentina).

Therefore, in Africa the oldest fossil of the *Madtsoia/Gigantophis* group comes from the Early Senonian, whereas in South America this group appears only in the Palaeocene. The fact that these large snakes are missing in South America prior to the level of Tiupampa (that is, about the Cretaceous–Cainozoic boundary) is probably significant, more than 20 Late Cretaceous localities having yielded herpetofaunas in South America (Estes and Baez 1985). Therefore, a dispersal from Africa to South America by the Late Cretaceous–Palaeocene should be considered. As for pipid frogs, the emerged Walvis and Rio Grande Rises could have acted as a land bridge.

Thus, the history of madtsoiid snakes seems very similar to that of pipid frogs: the family spread over Africa and South America at approximately the time of the separation of these two continents. Their dispersal may have taken place on the still conjoined block or across the incipient South Atlantic. Subsequently, an African group crossed the oceanic barrier by the Late Cretaceous–Early Palaeocene and reached South America: the Walvis and Rio Grande Rises probably provided a more or less continuous land route for this dispersal.

Crocodylia

A variety of crocodilians are known from the Upper Cretaceous of Africa and South America (Bonaparte 1978; Buffetaut 1982*a*, 1985*a*), but there are few families and no genera in common. Two groups of small terrestrial mesosuchians, the Libycosuchidae in Africa and the Notosuchidae in South America have been considered as the possible result of vicariant evolution from a common Early Cretaceous ancestral stock (possibly among the Uruguaysuchidae) on both sides of the proto-Atlantic (Buffetaut 1982*a*), but this hypothesis needs confirmation. One family of mesosuchians, the Trematochampsidae, is represented in the Upper Cretaceous of both Africa and South

America by closely related genera, about which the above-mentioned problem may be posed: does this indicate interchange across the proto-Atlantic, or slow vicariant evolution? Tremato-champsids were first described from the Lower Senonian of In Beceten (Niger) by Buffetaut (1974, 1976a), with the species *Trematochampsa taqueti*. In Africa, the trematochampsids seem to have persisted, with terrestrial ziphodont forms, until the Tertiary, with *Eremosuchus elkoholicus*, from the Lower Eocene of Algeria (Buffetaut 1989a), and they seem to have reached Europe as early as the Late Cretaceous (Buffetaut 1989b), there to give rise to ziphodont forms in the Early Tertiary (Buffetaut 1988a). The family is also known to occur in the Upper Cretaceous of Madagascar (Buffetaut and Taquet 1979b). The first trematochampsid to be reported from South America was *Itasuchus jesuinoi*, described by Price (1955) from the Late Cretaceous Bauru Group of Brazil as a possible goniopholidid. In 1985, it was shown that *Itasuchus* is in fact a tre-matochampsid closely related to *Tremato-champsa* (Buffetaut 1985b). Two possible explanations for the occurrence of very closely allied genera of non-marine crocodilians in Africa and in South America after these continents had become separated by a marine barrier were suggested (Buffetaut 1985b): interchange across the proto-Atlantic, either by swimming across what was still a relatively narrow seaway or through a possible temporary land connection (perhaps corresponding to one of the present sub-marine 'rises' of the South Atlantic), or slow vicariant evolution on each side of the proto-Atlantic. At the time, the lack of Early Cretaceous remains of trematochampsids made it difficult to choose between these hypotheses: the earliest known possible trematochampsid remain was then an isolated tooth from the Cenomanian of Egypt. Since then, trematochampsids have been found in South America in deposits older than the Late Albian separation of Africa from South America. In 1988, Chiappe described a new tre-matochampsid, *Amargasuchus minor*, on the basis of upper jaw material from the Hauterivian

La Amarga Formation of north western Patagonia. This is the oldest record of the family Trematochampsidae, and, as pointed out by Chiappe, it favours slow vicariant evolution as the cause for the occurrence of similar tremato-champsids in the Upper Cretaceous of Africa and South America. This interpretation is further sup-ported by the occurrence of another tremato-champsid in the Santana Formation of Brazil. In 1987, Kellner described a new crocodilian taxon, *Caririsuchus camposi*, on the basis of jaw frag-ments and a brief examination of a complete but unprepared specimen. The specimen was later examined in detail, after preparation, by one of us (EB); its skull characters indicate that it belongs to the family Trematochampsidae, and that it so closely resembles the Late Cretaceous *Itasuchus* that it should probably be included in that genus. In conclusion, it is likely that trematochampsids were present in both South America and Africa before those continents became separated, and it can be predicted that trematochampsid remains will eventually be found in the ante-Albian Creta-ceous of Africa.

In this connection, it may be mentioned that a Late Cretaceous and Early Tertiary family of mesosuchian crocodilians, the Dyrosauridae, is known to have occurred both in Africa and in South America (Buffetaut 1976b). However, the dyrosaurids are usually found in shallow marine coastal deposits, and there is no doubt that they were adapted to life in the sea, so that crossing the proto-Atlantic was probably easy for them (Buffetaut 1982a). Their geographical distribu-tion thus does not require dispersal along a land route. The same probably applies to the Gavia-lidae, which were present in both Africa and South America during the Tertiary: early gavials are often found in marine deposits, and they probably were able to cross the South Atlantic sometime in the Early Tertiary (Buffetaut 1982b). Today, the Crocodylidae occur in both Africa and South America, with different species of the genus *Crocodylus*; the biogeographical history of this genus is poorly known, but Tertiary dispersal across the Atlantic seems likely.

Dinosauria

Sauropoda

The dominant herbivorous dinosaurs in the Late Cretaceous of South America are the Titanosauridae (von Huene 1929; Powell 1986). As pointed out by Bonaparte (1986) and Bonaparte and Kielan-Jaworowska (1987), they seem to have played a similar important part on other Gondwanan continental blocks as well, including Africa, Madagascar, and India. It should be noted, however, that the Late Cretaceous record of titanosaurids in Africa is still scanty (Buffetaut 1988*b*), with the poorly known *Aegyptosaurus* from the Cenomanian of Baharija (Stromer 1932, 1936), specimens from the Maastrichtian of Mount Igdaman in Niger (Greigert *et al.* 1954), and remains from a newly discovered locality in Sudan (Buffetaut *et al.* 1990). Titanosaurids are also very abundant in the Upper Cretaceous of Europe (Buffetaut 1989*b*; Le Loeuff *et al.* 1989). Both the evolutionary history of the group and its biogeographical history are still poorly known, however, *Torniera*, from the Upper Jurassic of Tendaguru in East Africa, has been considered as the earliest known titanosaurid (von Huene 1956), but this is not accepted by all sauropod specialists. Titanosaurid vertebrae have been reported from the Wealden of England and from 'Middle Cretaceous' deposits in England and France (see Buffetaut 1989*c*), but the record of the family in the Lower Cretaceous of the southern continents is very scanty. Titanosaurids do not seem to have been reported from the Lower Cretaceous of South America, and their Early Cretaceous record in Africa seems limited to a few procoelous caudal vertebrae from several localities in Niger, described by Lapparent in 1960 (in 1976, Taquet mentioned the occurrence of titanosaurids at Gadoufaoua, but his comprehension of the family Titanosauridae is different from that currently accepted by most authors, since it includes the 'dicraeosaurines', to which the Gadoufaoua material is referred). Most of the material described by Lapparent comes from Mount Iguallala, a locality placed by Lapparent in the upper part of the 'Continental Intercalaire',

not far below the Cenomanian; these localities are probably Albian in age. In addition, a procoelous caudal vertebra has been reported by Lapparent (1960) from in Gall (Niger), a locality placed in the lower part of the Continental Intercalaire, presumably of Neocomian to Barremian age. Although the dominant sauropods in the Early Cretaceous of Africa seem to have been camarasaurids, as reported by Lapparent (1960) and as confirmed by recent British expeditions to Niger (A. C. Milner, personal communication), titanosaurids thus seem to have been present as well, and it is likely that they already had a vast Gondwanan distribution at the beginning of the Cretaceous. This, coupled with fairly sluggish subsequent vicariant evolution, would explain the occurrence of relatively similar forms in the Late Cretaceous of South America, Madagascar, India, Africa, and Europe. There is a priori no reason to assume a Late Cretaceous interchange of titanosaurids across the South Atlantic. However, only a better knowledge of the Late Cretaceous titanosaurids of Africa would permit detailed comparisons with South American forms and thus a better understanding of relationships between these sauropod assemblages.

Theropoda

Whereas tyrannosaurids were the dominant large theropods in the Late Cretaceous of mainland Asia and western North America, their counterparts on the Gondwanan continents and in Europe seem to have belonged to a different family, the Abelisauridae (Bonaparte 1986; Bonaparte and Kielan-Jaworowska 1987; Buffetaut *et al.* 1988; Buffetaut 1989*b*). Abelisaurids were first reported from the Maastrichtian of Patagonia by Bonaparte and Novas (1985), with *Abelisaurus comahuensis. Xenotarsosaurus bonapartei*, also of Late Cretaceous age, is based on postcranial material (Martinez *et al.* 1986). The earliest hitherto recorded abelisaurid is *Carnotaurus sastrei*, described by Bonaparte (1985) from the Albian of Patagonia. Outside South America, the Abelisauridae are probably represented in the Maastrichtian Lameta Formation of India by *Indosuchus matleyi* (Bonaparte 1986; Buffetaut *et al.* 1988). Their occurrence in Europe is attested by cranial and postcranial material (Buffetaut *et al.* 1988; Le Loeuff and Buffetaut in preparation). In Africa, little is known of Cretaceous theropods. Besides the very peculiar spinosaurids (see Buffetaut 1989*d*), large theropods are represented by the poorly known *Carcharodontosaurus saharicus*, first described (as *Megalosaurus*) on the basis of teeth from the Albian of Algeria by Depéret and Savornin (1927). Skull and postcranial elements of this form were described by Stromer (1931), who coined the generic name, from the Cenomanian of Baharija, and various teeth and bones from many Saharan localities were referred to it by Lapparent (1960 and Lavocat (1948, 1954). Also, Bouaziz *et al.* (1988) have referred isolated teeth from the Albian of southern Tunisia to *Carcharodontosaurus saharicus. Baharijiasaurus ingens*, also described from Baharija by Stromer (1934), is based on very incomplete material, lacking skull elements and teeth. Among the fragmentary theropod remains from Baharija described by Stromer, there is also a peculiar femur said to be reminiscent of *Erectopus sauvagei* Huene, a rather mysterious theropod from the Albian of France first described by Sauvage (1882). The abelisaurids seem to be characterized, among other features, by a peculiar femoral morphology (Le Loeuff and Buffetaut, in preparation), and some of the above mentioned African material may belong to this family. This would not be unexpected from a biogeographical point of view, since the known distribution of abelisaurids leads to the conclusion that they must have been present in Africa, but the available evidence is insufficient to allow definite conclusions.

CONCLUSIONS

The biogeographical history of various groups of amphibians and reptiles thus documents the consequences of the separation of Africa and South America during the Cretaceous. As evidenced by the close similarities between non-marine reptiles from the Aptian of Niger and Brazil, a fairly homogenous continental fauna was present in both Africa and South America until relatively late in the Early Cretaceous, and this points to the absence of any significant barrier between these land areas—which is in agreement with independent data indicating separation by a continuous seaway only in the Albian. Later faunas reveal increasingly greater differences between African and South American amphibian and reptile assemblages. The picture that emerges, however, is not one of simple vicariant evolution on each side of a widening oceanic barrier. There is convincing evidence that faunal interchange remained possible, to some extent, long after a continuous seaway had become established between Africa and South America in the Late Albian. At least among pipid frogs and madtsoiid snakes, groups which had evolved on the African continent were able to reach South America, even though they may not have flourished there for long. Rafting across the proto-Atlantic may account for some of these dispersal events, but the temporary existence of a filter bridge on the emplacement of the Walvis–Rio Grande Rises may be a better explanation of the transatlantic dispersal of animals such as frogs, known to be very sensitive to salt water. In any case, such dispersal events do not seem to have taken place after the Palaeocene.

A somewhat different picture is provided by crocodilians and dinosaurs. Several Early Cretaceous genera of crocodilians are represented by very closely allied, if not identical, species in Africa and in South America, prior to the establishment of a continuous proto-Atlantic seaway. The available evidence strongly suggests that the resemblances between the Late Cretaceous trematochampsids of Africa and South America may be due to slow vicariant evolution rather than to transatlantic dispersal. The same may be true of dinosaurs such as titanosaurids and abelisaurids, although the imperfect record of these groups in Africa makes definite conclusions difficult. In any case, if a temporary filter bridge became established at the Cretaceous–Tertiary boundary or close to it, perhaps during a period of low sea-level, it may have been too late for dinosaurs to use it.

REFERENCES

Ahl, E. (1926). Anura; Aglossa, Xenopodidae. In *Die Diamantenwüste Südwest-Aflrikas*, Vol. 2, (ed. E. Kaiser), pp. 141–2. D. Reimer, Berlin.

Albino, A. M. (1986). Nuevos Boidae Madtsoiinae en el Cretácico tardío de Patagonia (Formación Los Alamitos, Rio Negro, Argentina). IV *Congreso Argentino de Paleontologia y Bioestratigrafia, Actas*, vol. 2, pp. 15–21, Mendoza.

Allport, S. (1860). On the discovery of some fossil remains near Bahia in South America. *Quarterly Journal of the Geological Society of London*, **16**, 263–8.

Baez, A. M. (1981). Redescription and relationships of *Saltenia ibanezi*, a late Cretaceous pipid frog from northwestern Argentina, *Ameghiniana*, **18**, 127–54.

Baez, A. M. and Rage, J. C. (1988). Evolutionary relationships of a new pipid frog from the Upper Cretaceous of Niger. In *Herpetologists League Meeting*, p. 16. Ann Arbor.

Bisbee, C. A., Baker, M. A., Wilson, A. C., Hadji-Azimi, I., and Fischberg, M. (1977). Albumin Phylogeny for clawed frogs *(Xenopus), Science*, **195**, 785–7.

Bonaparte, J .F. (1978). *El Mesozoico de America del Sur y sus Tetrapodos.* Opera Lilloana 26, San Miguel de Tucuman.

Bonaparte, J. F. (1985). A horned Cretaceous carnosaur from Patagonia. *National Geographic Research*, **1**, 149–52.

Bonaparte, J. F. (1986). History of the terrestrial Cretaceous vertebrates of Gondwana. In IV *Congreso Argentino de Paleontologia y Bioestratigrafia, Actas*, Vol. 2, pp. 63–95, Mendoza.

Bonaparte, J. F. and Kielan-Jaworowska, Z. (1987). Late Cretaceous dinosaur and mammal faunas of Laurasia and Gondwana. In *Fourth Symposium on Mesozoic Terrestrial Ecosystems, Short Papers* (ed. P. J. Currie and E. H. Koster), pp. 24–9, Tyrrell Museum of Palaeontology, Drumheller.

Bonaparte, J. F. and Novas, F. E. (1985). *Abelisaurus comahuensis*, n.g., n.sp., Carnosauria del Cretacico tardio de Patagonia. *Ameghiniana*, **21**, 259–65.

Bouaziz, S., Buffetaut, E., Ghanmi, M., Jaeger, J. J., Martin, M., Mazin, J. M., and Tong, H. (1988). Nouvelles découvertes de vertébrés fossiles dans l'Albien du Sud tunisien. *Bulletin de la Société géologique de France*, **4**, 335–9.

Brito, I. M. and Campos, D. A. (1983). The Brazilian Cretaceous. *Zitteliana*, **10**, 227–83.

Broin, F. de (1980). Les tortues de Gadoufaoua (Aptien du Niger); aperçu sur la paléobiogéographie des Pelomedusidae (Pleurodira). *Mémoires de la Société géologique de France*, **138**, 39–46.

Broin, F. de (1988). Les tortues et le Gondwana. *Studia palaeocheloniologica*, **2**, 103–42.

Broin, F. de and Taquet, P. (1966). Découverte d'un Crocodilien nouveau dans le Crétacé inférieur du Sahara. *Comptes rendus de l'Académie des Sciences de Paris*, D, **262**, 2326–9.

Broin, F. de, Buffetaut, E., Koeniguer, J. C., Rage, J. C., Russell, D., Taquet, P., Vergnaud-Grazzini, C., and Wenz, S. (1974). La faune de Vertébrés continentaux du gisement d'In Beceten (Sénonien du Niger). *Comptes rendus de l'Académie des Sciences de Paris*, D, **279**, 469–72.

Buffetaut, E. (1974). *Trematochampsa taqueti*, un crocodilien nouveau du Sénonien inférieur du Niger. *Comptes rendus de l'Académie des Sciences de Paris*, D, **279**, 1749–52.

Buffetaut, E. (1976*a*). Ostéologie et affinités de *Trematochampsa taqueti* (Crocodylia, Mesosuchia) du Sénonien inférieur d'In Beceten (République du Niger). *Geobios*, **9**, (2), 143–98.

Buffetaut, E. (1976*b*). Sur la répartition géographique hors d'Afrique des Dyrosauridae, crocodiliens mésosuchiens du Crétacé terminal et du Paléogène. *Comptes rendus de l'Académie des Sciences de Paris*, D, **283**, 487–90.

Buffetaut, E. (1981), Die biogeographische Geschichte der Krokodilier, mit Beschreibung einer neuen Art, *Araripesuchus wegeneri. Geologische Rundschau*, **70**, (2), 611–24.

Buffetaut, E. (1982*a*). Radiation évolutive, paléoécologie et biogéographie des crocodiliens mésosuchiens. *Mémoires de la Société géologique de France*, **80**, (142), 1–88.

Buffetaut, E. (1982*b*). Systématique, origine et évolution des Gavialidae sud-américains. *Geobios Mémoire spécial*, **6**, 127–40.

Buffetaut, E. (1985*a*). Zoogeographical history of African crocodilians since the Triassic. In *Proceedings of the International Symposium on African vertebrates* (ed. K. H. Schuchmann), pp. 453–69, Museum Alexander Koenig, Bonn.

Buffetaut, E. (1985*b*). Présence de Trematochampsidae (Crocodylia, Mesosuchia) dans le Crétacé supérieur du Brésil. Implications paléobiogéographiques. *Comptes rendus de l'Académie des Sciences de Paris*, II, **301**, 1221-4.

Buffetaut, E. (1988*a*). The ziphodont mesosuchian crocodile from Messel: a reassessment. *Courier Forschungsinstitut Senckenberg*, **107**, 211–21.

Buffetaut, E. (1988*b*). Late Cretaceous sauropod dinosaurs of Africa: a comment. *South African Journal of Science*, **84**, 221.

Buffetaut, E. (1989*a*). A new ziphodont mesosuchian crocodile from the Eocene of Algeria. *Palaeontographica* A, 208, 1–10.

Buffetaut, E. (1989*c*). Archosaurian reptiles with Gondwanan affinities in the Upper Cretaceous of Europe. *Terra nova*, 1, 69-74.

Buffetaut, E. (1989*c*). Une vertèbre de dinosaure titanosauridé dans le Cénomanien du Mans et ses implications paléobiogéographiques. *Comptes rendus de l'Académie des Sciences de Paris*, II, **309**, 437–443.

Buffetaut, E. and Taquet, P. (1976). Le crocodilien géant *Sarcosuchus* dans le Crétacé inférieur du Brésil (bassin de Bahia) et du Niger (bassin du Tegama). In *4e Réunion Annuelle des Sciences de la Terre*, Société géologique de France, Paris, p. 82.

Buffetaut, E. and Taquet, P. (1977*a*). The giant crocodilian *Sarcosuchus* in the Early Cretaceous of Brazil and Niger. *Palaeontology*, **20**, 203–8.

Buffetaut, E. and Taquet, P. (1977*b*). Un crocodile géant à cheval sur deux continents. *La Recherche*, **76**, 289–91.

Buffetaut, E. and Taquet, P. (1979*a*). An Early Cretaceous terrestrial crocodilian and the opening of the South Atlantic. *Nature*, **280**, 486–7.

Buffetaut, E. and Taquet, P. (1979*b*). Un nouveau crocodilien mésosuchien dans le Campanien de Madagascar: *Trematochampsa oblita*, n.sp. *Bulletin de la Société géologique de France*, **21**, 183–8.

Buffetaut, E., Mechin, P., and Mechin-Salessy, A. (1988). Un dinosaure théropode d'affinités gondwaniennes dans le Crétacé supérieur de Provence. *Comptes rendus de l'Académie des Sciences de Paris*, II, **306**, 153–8.

Buffetaut, E., Bussert, R., and Brinkmann, W. (1990). A nonmarine vertebrate fauna in the Upper Cretaceous of northern Sudan. *Berliner geowissenschaftliche Abhandlungen*, A, **120**, 183–202.

Cannatella, D. C. and Trueb, L. (1988*a*). Evolution of pipoid frogs: intergeneric relationships of the aquatic frog family Pipidae (Anura). *Zoological Journal of the Linnean Society*, **94**, 1-38.

Cannatella, D. C. and Trueb, L. (1988*b*). Evolution of pipoid frogs: morphology and phylogenetic relationships of *Pseudhymenochirus*. *Journal of Herpetology*, **22**, 439–56.

Chiappe, L. M. (1988). A new trematochampsid crocodile from the early Cretaceous of north-western Patagonia, Argentina and its palaeobiogeographical and phylogenetic implications. *Cretaceous Research*, **9**, 379–89.

Cuny, G., Jaeger, J. J., Mahboubi, M., and Rage, J. C. (1990). Les plus anciens serpents (Reptilia, Squamata) connus. Mise au point sur l'âge géologique des serpents de la partie moyenne du Crétacé. *Comptes rendus de l'Académie des Sciences de Paris*. II, **311**, 1267–72.)

Depéret, C. and Savornin, J. (1927). La faune de Reptiles et de Poissons albiens de Timimoun (Sahara occidental). *Bulletin de la Société géologique de France*, **27**, 257–65.

Estes, R. (1975). Fossil *Xenopus* from the Paleocene of South America and the zoogeography of pipid frogs. *Herpetologica*, **31**, 263–78.

Estes, R. (1977). Relationships of the South African fossil frog *Eoxenopoides reuningi* (Anura, Pipidae). *Annals of the South African Museum*, **73**, 49–80.

Estes, R. and Baez, A. (1985). Herpetofaunas of North and South America during the late Cretaceous and Cenozoic: Evidence for interchange? In *The Great American Biotic Interchange* (ed. F. G. Stehli and S. D. Webb), pp. 139–97. Plenum Press, New York.

Estes, R., Špinar, Z. V., and Nevo, E. (1978). Early cretaceous pipid tadpoles from Israel (Amphibia: Anura). *Herpetologica*, **34**, 374–93.

Förster, R. (1978). Evidence for an open seaway between northern and southern proto-Atlantic in Albian times. *Nature*, **272**, 158–9.

Gheerbrant, E. (1987). Les vertébrés continentaux de l'Adrar Mgorn (Maroc, Paléocène); une dispersion de mammifères transtéthysienne aux environs de la limite Mésozoïque/Cénozoïque? *Geodinamica Acta*, **1**, 233–46.

Greigert, J., Joulia, F., and Lapparent, A. F. de (1954). Répartition stratigraphique des gisements de Vertébrés dans le Crétacé du Niger. *Comptes rendus de l'Académie des Sciences de Paris*, D, **239**, 433–5.

Huene, F. von (1929). Los saurisquios y ornitisquios

del Cretacico argentino, *Anales del Museo de La Plata*, **3**, 1–196.

Huene, F. von (1956). *Paläontologie und Phylogenie der niederen Tetrapoden*. Gustav Fischer Verlag, Jena.

Kellner, A. W. A. (1987). Ocorrência de um novo crocodiliano no Cretaceo inferior da Bacia do Araripe, Nordeste do Brasil. *Anais da Academia Brasileira de Ciências*, **59**, 220–32.

Krömmelbein, K. (1971). Non-marine Cretaceous ostracodes and their importance for the hypothesis of Gondwanaland. *Proceedings of the 2nd IUGS Gondwana Symposium*, Pretoria pp. 617–19.

Lapparent, A. F. de (1960). Les dinosauriens du 'Continental Intercalaire' du Sahara central. *Mémoires de la Société géologique de France*, n.s., **39**, 88A, 1–57.

Lavocat, R. (1948). Découverte de Crétacé à Vertébrés dans le soubassement de la Hammada du Guir (Sud marocain). *Comptes rendus de l'Académie des Sciences de Paris*, D, **226**, 1291–2.

Lavocat, R. (1954). Sur les dinosauriens du Continental intercalaire des Kem-Kem de la Daoura. In *19ème Congrès géologique international*, Alger 1952, Section XIII–3, *fasc. XV*, pp. 65–8.

Le Loeuff, J., Buffetaut, E., Mechin, P., and Mechin-Salessy, A. (1989). Un arrière-crâne de dinosaure titanosauridé (Saurischia, Sauropoda) dans le Crétacé supérieur du Var (Provence, France). *Comptes rendus de l'Académie des Sciences de Paris*, II, **309**, 851–7.

Marsh, O. C. (1869). Notice on some new reptilian remains from the Cretaceous of Brazil. *American Journal of Science*, **47**, 390-2.

Martinez, R., Giménez, O., Rodriguez, J., and Bochatey, G. (1986). *Xenotarsosaurus bonapartei* nov. gen. et sp. (Carnosauria, Abelisauridae), un nuevo Theropoda de la Formacion Bajo Barreal, Chubut, Argentina. In IV *Congreso Argentino de Paleontologia y Bioestratigrafia. Actas*, vol. 2, pp. 23–31, Mendoza.

Mawson, J. and Woodward, A. S. (1907). On the Cretaceous formation of Bahia (Brazil) and on vertebrate fossils collected therein. *Quarterly Journal of the Geological Society of London*, **63**, 128-39.

Mimran, Y. (1972). The Tayasir Volcanics, a Lower Cretaceous formation in the Shomeron, central Israel. *Israel Geological Survey Bulletin*, **52**, 1–9.

Nevo, E. (1968). Pipid frogs from the early Cretaceous of Israel and pipid evolution. *Bulletin of the Museum of Comparative Zoology*, **136**, 255–318.

Powell, J. E. (1986). Revision de los Titanosauridos de America del Sur. Unpublished thesis, Universidad Nacional de Tucuman.

Price, L. I. (1955). Novos crocodilideos dos arenitos da Seria Bauru, Cretaceo do Estado de Minas Gerais. *Anais da Academia brasileira de ciências*, **27**, 487-98.

Price, L. I. (1959). Sôbre um Crocodilideo Notossuquio do Cretacico Brasileiro. *Divisao de Geologia e mineralogia, Boletim*, **188**, 7–55.

Price, L. I. (1973). Quelônio Amphichelydia no Cretaceo inferior do Nordeste do Brasil. *Revista Brasileira de Geociencias*, **3**, 84–96.

Rage, J. C. (1984). Serpentes. *Handbuch der Paläoherpetologie* (ed. P. Wellnhofer), 11, xii + 80 p. G. Fischer, Stuttgart.

Rage, J. C. (1987). Fossil History. In *Snakes, Ecology and evolutionary biology* (ed. R. Seigel, J. T. Collins, and S. S. Novak), pp. 51–76. McMillan, New York.

Rage, J. C. (1988). Gondwana, Tethys and terrestrial vertebrates during the Mesozoic and Cainozoic. In *Gondwana and Tethys* (ed. M. G. Audley-Charles and A. Hallam), pp. 255–73. Geological Society special paper no. 37. Oxford University Press.

Rage, J. C. (1991). Squamate Reptiles from the early Paleocene of the Tiupampa area (Santa Lucía Formation), Bolivia. *Revista Tecnica de Yacimientos petroliferos fiscales Bolivianos*, **12** (3–4), 503–8.

Reyment, R. (1980). Paleo-oceanology and the paleobiogeography of the Cretaceous South Atlantic Ocean. *Oceanologica Acta*, **3**, 127-33.

Sauvage, H. E. (1882). Recherches sur les reptiles trouvés dans le Gault de l'Est du Bassin de Paris. *Mémoires de la Société géologique de France*, **2**, 1–43.

Stromer, E. (1931). Ergebnisse der Forschungsreisen Prof. E. Stromers in den Wüsten Ägyptens. II. Wirbeltier-Reste der Baharîje-Stufe (unterstes Cenoman). 10. Ein Skelettrest von *Carcharodontosaurus* nov. gen. *Abhandlungen der Bayerischen Akademie der Wissenschaften*, N.F., **9**, 1–23.

Stromer, E. (1932). Ergebnisse der Forschungsreisen Prof. E. Stromers in den Wüsten Ägyptens. II. Wirbeltier-Reste der Baharîje-Stufe (unterstes Cenoman). 11. Sauropoda. *Abhandlungen der Bayerischen Akademie der Wissenschaften*, N.F., **10**, 1–21.

Stromer, E. (1934). Ergebnisse der Forschungsreisen Prof. E. Stromers in den Wüsten Ägyptens. II. Wirbeltier-Reste der Baharîje-Stufe (unterstes Cenoman). 13. Dinosauria. *Abhandlungen der*

Bayerischen Akademie der Wissenschaften, N.F., **22**, 1–79.

Stromer, E. (1936). Ergebnisse der Forschungsreisen Prof. E. Stromers in den Wüsten Ägyptens. VII. Baharîje-Kessel und Stufe mit deren Fauna und Flora. Eine ergänzende Zusammenfassung. *Abhandlungen der Bayerischen Akademie der Wissenschaften*, N.F., **33**, 1–102.

Taquet, P. (1976). Géologie et paléontologie du gisement de Gadoufaoua (Aptien du Niger). *Cahiers de Paléontologie*, **1976**, 1–188.

Vergnaud-Grazzini, C. (1966). Les Amphibiens du Miocène de Beni Mellal. *Notes du Service géologique du Maroc*, **27**, 43-75.

Wenz, S. (1980). A propos du genre *Mawsonia*, coelacanthe géant du Crétacé inférieur d'Afrique et du Brésil. *Mémoires de la Société géologique de France*, **139**, 187–90.

8 THE ORIGIN OF THE NEW WORLD MONKEYS

Leslie C. Aiello

INTRODUCTION

The New World or platyrrhine, monkeys are found today in the tropical regions of both Central and South America. Their fossil record extends back approximately 26 million years to the Late Oligocene Deseadan Beds of Bolivia (*Branisella boliviana*) (MacFadden 1990), and this poses a mystery. Throughout most of the early Cenozoic South America was an island continent (Tarling 1980). If primates arrived in South America in the Late Eocene or Early Oligocene as is commonly thought, they would have had to cross a considerable water barrier no matter where they came from.

Because there is no obvious route of entry for the New World monkeys into South America there is considerable controversy over their evolutionary ancestry. Early authors, who accepted as a basic premise the static position of the continents and seas, argued for the immediate ancestry of the platyrrhine primates among the Eocene pre-anthropoid primates of North America (Loomis 1911; Gregory 1920, 1921; Gazin 1958; Simons 1961). The ancestry of the catarrhine primates of the Old World was sought among the Eocene pre-anthropoid primates of Eurasia. Because of the wide oceanic barriers currently separating the Americas from Eurasia and Africa, the morphological similarity between the platyrrhine and catarrhine primates was necessarily, and sometimes reluctantly, explained as a parallel development from similar pre-anthropoid ancestors.

In the more recent literature few authors have continued to support an explanation for New World monkey origins that requires parallel evolution of the New and Old World primates from separate non-anthropoid ancestors (but see

Simons 1972; Cachel 1979). The acceptance of plate tectonics has radically altered the conception of the geography of the early Cenozoic and made the connection conceivable between groups of animals that would have been implausible, if not impossible, if the geography of that time had been identical to that found today. In addition, intensive work in the areas of biochemistry and comparative morphology have underscored the similarities between the platyrrhine and catarrhine primates and added support to the hypothesis that these primates shared a unique common ancestor (see Delson and Rosenberger 1980; Harrison 1987; Martin 1990*a* for summaries of the evidence).

There are currently two major and rival schools of thought concerning the origin of the New World primates that are based on the monophyly of the platyrrhine and catarrhine primates and on modern ideas of plate tectonics. These rival schools have been labelled the Northern Continents Model (Fig. 8.1) and the Southern Continents model (Fig. 8.2) (Delson and Rosenberger 1980). The Northern Continents Model argues that the ancestors of the higher, or anthropoid, primates (monkeys, apes, and humans) first appeared in the northern hemisphere. The New World monkeys then entered South America sometime in the Late Eocene or earliest Oligocene by waif dispersal across the ancestral Caribbean (Gingerich 1980; Delson and Rosenberger 1980). The Old World primates, or catarrhine primates, would have arrived in Africa at about the same time by dispersing across the Tethys Sea which separated Africa from Europe and Asia.

The alternative Southern Continents Model argues for the immediate ancestry of the New World primates in the southern continents (Hoffstetter 1972, 1974, 1980; Hershkovitz 1972,

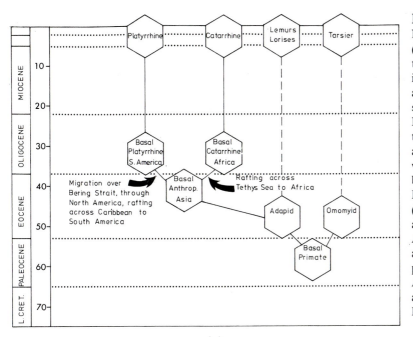

(a)

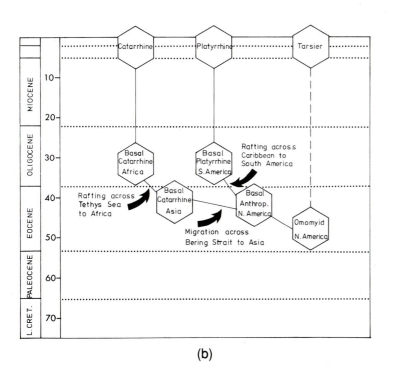

(b)

Fig. 8.1 Two versions of the Northern Continents Model. (a) Gingerich (1980) proposes that the basal anthropoid arose in Asia from adapid ancestors and that the basal platyrrhine spread over the Bering Strait to North America, migrated through North America, and arrived in South America via waif dispersal across the Caribbean in the Late Eocene. (b) Delson and Rosenberger (1980) propose that the basal anthropoid arose in North America from an omomyid ancestor, and that the basal platyrrhine arrived in South America via waif dispersal across the Caribbean in the Late Eocene.

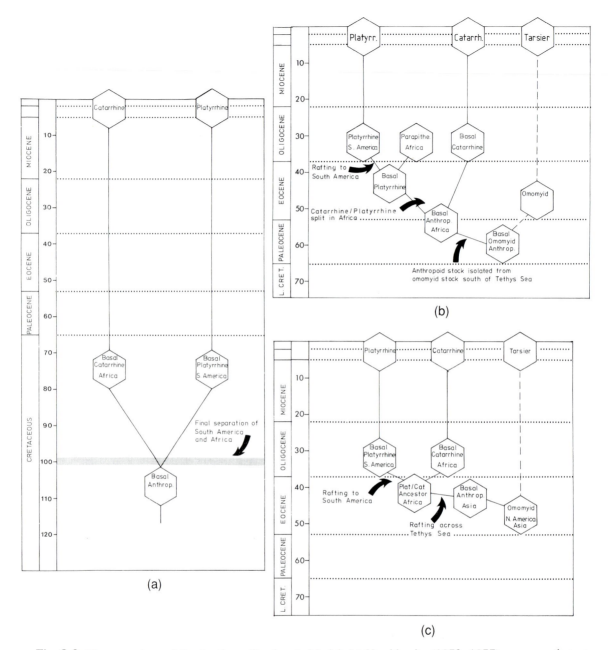

Fig. 8.2 Three versions of the Southern Continents Model. (a) Hershkovitz (1972, 1977) proposes that an unknown basal anthropoid stock was present on the combined southern continents prior to their complete separation. The division of this basal stock into the catarrhines and the platyrrhines was initiated by the formation of the South Atlantic ocean (ca 100 mya). (b) Hoffstetter (1972, 1974, 1980) proposes that an unknown basal anthropoid stock occupied Africa in the Palaeocene, that the catarrhine/platyrrhine split occurred on the African continent in the Late Palaeocene or Early Eocene, and that the platyrrhines arrived in South America via waif dispersal across the South Atlantic in the Late Eocene. (c) Ciochon and Chiarelli (1980) suggest that the basal anthropoid evolved in the northern hemisphere while the platyrrhines had their immediate ancestry on the African continent.

1977; Cracraft 1974; Lavocat 1974, 1977, 1980; Ciochon and Chiarelli 1980; Conroy 1980; Luckett 1980; Sarich and Cronin 1980; Fleagle and Kay 1987; Fleagle 1988).

There are currently three versions of the Southern Continents Origin Model. The first version proposes that an unknown basal anthropoid stock was present on the united southern continents prior to the complete separation of Africa and South America in the Cretaceous (about 100 my ago). The evolution of the New and Old World primates from this basal anthropoid stock is seen as the result of the separation of the continents resulting in the genetic isolation of the descendant populations.

The second version of the Second Southern Continents Model argues that anthropoid primates were present in Africa in the early part of the Cenozoic prior to the earliest Oligocene, but subsequent to the separation of the southern continents. This version proposes that the platyrrhines reached South America from Africa by waif dispersal across a relatively narrow Palaeocene, Eocene or Oligocene South Atlantic Ocean (Hoffstetter 1972, 1974, 1980, 1988; Fleagle 1988).

The third version argues that the anthropoid primates evolved in the northern continents during the Eocene and then dispersed across the Tethys Sea to Africa in the Late Eocene. The platyrrhines ultimately reached South America from Africa sometime in the Oligocene (Ciochon and Chiarelli 1980).

Two of these hypotheses, the Northern Continents Model and the third version of the Southern Continents Model, are based on the assumption that primates first evolved in the northern hemisphere and were confined to North America and Eurasia during the Palaeocene and Eocene periods. The other two versions of the Southern Continents Model argue that primates were present in the southern hemisphere during the early Cenozoic (or even earlier).

The following assessment of the relative merits of the various ideas for the origin of the New World monkeys will begin with a discussion of two related questions. Firstly, it will assess the probability that the distribution of Palaeocene and Eocene primates included the southern continents and, secondly, it will address the possibility that primates could have been present on the united southern continents prior to their separation in the later part of the Cretaceious period. Following on from these discussions, the Northern and Southern Continents Models will be evaluated in terms of:

1. the probability of dispersal across the South Atlantic Ocean in the early Cenozoic

2. the question of the immediate ancestry of the New World monkeys among some of the earliest anthropoid primates from Africa, the parapithecids of the Fayum in Egypt

3. the significance of other possible platyrrhine ancestors found elsewhere in the world.

THE DISTRIBUTION OF PRIMATES IN THE PALAEOCENE AND EOCENE

The majority of pre-Oligocene primates can be divided into two major grades, or stages, of evolution. The earliest of these stages included primates from the very Late Cretaceous and the Palaeocene periods that belong to the suborder Plesiadapiformes. Except for the earliest of these (*Purgatorius*), all of the Plesiadapiformes are dentally so specialized that they cannot reasonably be considered as direct ancestors of any later occurring primates (Szalay and Delson 1979; Martin 1990*a*). Because of their unique specialized morphology they are generally known as archaic primates (Martin 1990*a*) and some authors are in favour of excluding them from the order Primates altogether (Wible and Covert 1987). It is even suggested that at least some of these archaic 'primates' may be more closely related to the flying lemurs or colugos (Dermoptera) than they are to the primates (Beard 1990; Kay *et al.* 1990; Martin 1990*b*).

The second stage of early primate evolution is largely known from the Eocene period although one newly discovered representative is of Palaeocene age (Sigé *et al.* 1990), and some late occurring fossils extend into the Miocene period (Gingerich 1979; Gingerich and Sahni 1984; Wu and Pan 1985; Pan and Wu 1986). These second stage primates belong to two families, the Adapidae and the Omomyidae, and are similar in their general grade, or stage, of evolution to modern lemurs, lorises, and tarsiers. Because of their similarity to modern primates they are often called 'euprimates' and are generally taken to represent the first and earliest appearance of primates of modern form.

Stage three primates, or ancestral anthropoid (higher) primates, are distinguished from the Adapidae and the Omomyidae by the presence of an orbital plate, a fused metopic suture, a fused mandibular symphysis, and details of the dentition. There is some arguable evidence that stage three primates first appeared in eastern Asia in the Late Eocene period (Maw *et al.* 1979). However, more recent evidence suggests that they first appeared in North Africa in the Early or Middle Eocene (Godinot and Mahboubi 1992), and were well established in Africa and the Arabian peninsula by the late Eocene and Early Oligocene (Fleagle *et al.* 1986), Pickford 1986, Simons 1990, Thomas *et al.* 1991, Van Couvering and Harris 1991). They are also present in the Late Oligocene at Salla in Bolivia, South America (MacFadden 1990).

The traditional interpretation of this fossil record suggests that euprimates first evolved in North America or Europe at the beginning of the Eocene period and that anthropoid primates would have evolved in the Late Eocene in Asia. Primates would not have dispersed to Africa until the latest Eocene and to South America until the Late Oligocene.

The major problem with this reasoning is the belief that the fossil record offers anything more than the merest, and in some cases highly distorted, glimpse of the course of primate evolution (see Martin 1990*a*). This is illustrated by the recent discovery of *Altiatlasius koulchii*, an omomyid (euprimate) in Palaeocene deposits in North Africa (Sigé *et al.*, 1990). This new find suggests that euprimates may have in fact first evolved in Africa (see also Gingerich 1986).

In North America the euprimates are part of the Wasatchian Land Mammal Age (Rose 1980) and in Europe they are part of the Sparnacian Faunal Age (Gingerich 1981). Both the Wasatchian and the Sparnacian are largely immigrant faunas, with relatively few species in either which are found in earlier levels in North America or Europe (Savage and Russell 1983, Krause and Maas 1990). In the Wasatchian of North America the euprimates (omomyids and adapids) are accompanied by the first appearance of Artiodactyla (*Diacodexis*, Perissodactyla (*Hyracotherium*) and hyaenodontid creodonts. In the Sparnacian of Europe as much as 85 per cent of the fauna is believed to represent immigrants into the area. There is a large degree of faunal correspondence between the Wasatchian and Sparnacian. About 50 per cent of the Sparnacian genera are also found in the Wasatchian. This is the largest degree of faunal correspondence found at any period in the Cenozoic between two major continental areas and suggests a broad terrestrial connection between Europe and North America in the latest Palaeocene and the Early Eocene (Savage and Russell 1983).

The appearance of both the Sparnacian and the Wasatchian faunas correlates with a marked improvement in the climate (Gingerich 1977, 1980). Throughout the preceding Palaeocene the climate in North America and Europe deteriorated to a point where the Late Palaeocene was significantly cooler than the subtropical conditions of the Early and Middle Palaeocene (Wolfe 1975, 1978, 1980). By the beginning of the Eocene and the appearance of the Sparnacian and Wasatchian faunas the climate had warmed again and become subtropical. The Wasatchian fauna could have originated in Central America and, following the northward shift of the subtropical ecological zone, entered North America and then spread from there across the North Atlantic 'land bridge' to Europe (Sloan 1969). However, there is also evidence of faunal similarity between eastern

Asia and both the Wasatchian and Sparnacian faunas (Gingerich and Rose 1977; Rose 1980). This similarity suggests a possible Holarctic distribution of Late Palaeocene and Early Eocene faunas that would have been made possible not only by an open North Atlantic Land Bridge (McKenna 1983) but also by an open Bering Straits Land Bridge (Wolfe 1975) between eastern Asian and North America. Dispersal would have been facilitated by tropical land subtropical climates in the more northern latitudes (Wolfe 1975, 1978; McKenna 1980).

Mammalian faunas from Mongolia provide support for this hypothesis in that they are similar to both the Wasatchian as well as the immediately preceding Clarkforkian Mammalian Ages of North America (Rose 1980). In addition, *Altanius orlovi* confirms the existence of a primate in eastern Asia at a time contemporary with (or possibly slightly earlier than) the first appearance of the euprimates (adapids and omomyids) in North America (Dashzeveg and McKenna 1977). *Altanius* shows its closest morphological similarity to the anaptomorphine omomyids of the Wasatchian (*Anemorhysis*, *Tetonius*). However it also has some phenetic similarity to the plesiadapids (Rose and Krause 1984) as well as to the early notharctine (adapid) *Pelycodus* (Wasatchian) (Szalay and Delson 1979). Gingerich (1990) emphasizes that both *Altanius* and the newly discovered *Altiatlasius koulchii* from the North African Palaeocene are both so primitive as to not fit clearly into either familial taxon.

The idea that the euprimates entered North America at the beginning of the Eocene over the Bering Straits presupposes that primates were present in eastern Asia earlier than *Altanius*. Non-primate fossils associated with the North African Palaeocene omomyid primate *Altiatlasius* are also found in both Europe and North America, suggesting that at least intermittant communication with the northern continents was possible (Gingerich 1990).

Supporters of the first two versions of the Southern Continents Model (Figs 8.2a and b) would argue that euprimates occupied Africa during the Palaeocene climate deterioration and some would even argue that these primates would have evolved in Africa. With climatic improvement in the Early Eocene, these subtropical euprimates would have moved north, assuming a Holarctic distribution aided by both an equitable climate in the northern latitudes and the open Bering Strait and North Atlantic land bridges (see also Krause and Maas 1990).

In addition to the discovery of *Altialtasius koulchii*, the new Palaeocene North African primate, some of the best evidence for a long period of primate occupation in Africa is the diversity of the primate fauna in the Jebel Qatrani Formation of the Egyptian Fayum and in the Ashawq in Oman (Thomas *et al.* 1991). The argument is simply that the diversity of primates found in these Areas would have required a relatively long period of time to develop (Cooke 1968, 1972; Simons 1972). The primates bearing levels in Oman are of the Early Oligocene, while those at the Fayum are older than 31 my (Fleagle *et al.* 1986). It has recently been suggested that the entire Fayum Jebel Qatrani Formation may be Late Eocene (van Couvering and Harris 1991). At present there are 15 separate primate species recognized at the Fayum, 13 of which are of the anthropoid grade of evolution (Table 8.1) (Simons 1989, 1990). Of the other two, a single tooth documents the earliest occurrence of a loris in Africa (Simons *et al.* 1987) and a fragmentary jaw (*Afrotarsius chatrathi*) the first and only occurrence of an African tarsioid primate (Simons and Bown 1985), The remaining 13 species are divided into two higher primate families. Five of them belong to the Parapithecidae, the more primitive of the two families. The parapithecids are the only Old World anthropoids to retain three premolars and they also retain a number of primitive cranial and dental features, some of which are also found in the New World monkeys (Fleagle and Kay 1987; Fleagle 1988). The remaining species belong to the generally more advanced Propliopithecidae. These primates have only two premolars as do all of the later occurring Old World anthropoid primates.

Table 8.1 The higher primates from the Jebel Qatrani Formation of the Egyptian Fayum (Simons 1989, 1990)

Parapithecidae
  Qatrania wingi
  Apidium phiomense
  Apidium moustafai
  Parapithecus fraasi
  Parapithecus grangeri

Propliopithecidae
  Oligopithecinae
    Catopithecus browni
    Oligopithecus savagei
    Proteopithecus sylviae
  Propliopithecinae
    Propliopithecus haeckeli
    Propliopithecus chirobates
    Propliopithecus markgrafi
    Propliopithecus ankeli
    Aegyptopithecus zeuxis

The earliest and most primitive of these propliopithecids, the Oligopithecinae, have dental features that have been interpreted as linking these early anthropoid primates with the Eocene adapids (Simons 1990). The later and more advanced Propliopithecinae are dentally more similar to living apes but are still primitive enough to most probably have preceded the division of the Old World monkeys and the apes.

The similarity between the earlier Fayum anthropoids and the more primitive euprimates of the Eocene suggests that the anthropoids most probably evolved in Africa (Fleagle 1988; Simons 1990). The antiquity of anthropoid primates in Africa is also consistent with the evidence which suggests that the platyrrhine and the catarrhine primates had a long period of common ancestry prior to their separation into independent lines. At the most extreme, Sarich (1970) has suggested, on the basis of immunological distances, that they shared a period of common ancestry lasting 30 million years. While this conclusion can be questioned on methodological grounds (Baba *et al.* 1980; and references therein), Ciochon and Chiarelli (1980) emphasize that the inferred shared derived features common to the New World monkeys and the Old World monkeys are twice as numerous as, and are more significant in terms of level of organization (structural functional changes) than, the inferred derived features characterizing either the New World monkeys or the Old World monkeys.

Other support for the possible existence of primates in Africa prior to the Oligocene comes from the evidence of the extant fauna of Madagascar which suggests that primates of modern aspect could have occupied Africa as long ago as the latest Cretaceous. Tattersall (1982) has argued

... that there is a high probability that the African mammalian fauna in the early Tertiary both resembled that of Madagascar and was substantially different in composition and diversity from that represented in the subsequent fossil records.

In Tattersall's view this Early Tertiary fauna would have included tethytheres (proboscidians, hydracoids, sirenians, etc.), tenrecs, primates, and possibly condylarths of which small mammals, such as the tenrecs and primates, together with non-mammals, such as the ratite birds, survived into subrecent or recent times on Madagascar.

Evidence to support this revolves around two major points. Firstly, the Fayum mammal fauna, as well as subsequent African faunas, show a component of European forms which the earlier Africa faunas, as well as the Madagascar fauna, do not (Savage and Russell 1983). The Early Tertiary African fauna (based on the admittedly limited fossil record) may well have resembled the modern Madagascar fauna in being characterized by a low diversity of major taxa associated with a high diversity within these taxa (Tattersall 1982). Secondly, Tattersall suggests that the inferred phylogenetic affinities of the strepsirhine primates (lemurs, lorises, adapids, and omomyids) indicate that a large part of the diversification of the lemur fauna might have taken place before the complete isolation of Madagascar from Africa. In particular, he argues that the Madagascar lemurs do not form a unique clade within the strepsirhines as would be expected if they originated from a single basal lemur that arrived in Madagascar as the result of one event of waif dispersal. Rather, Tattersall suggests that the cheirogaleids (the Dwarf Lemurs of Madagascar) are the sister group of the lorises of mainland Africa as well as Asia and, therefore, are most closely related to them and not to the remaining Malagasy lemurs (also supported by Szalay and Katz 1973; Tattersall and Schwartz 1974, 1975; Hoffstetter 1974; and Cartmill 1975; but see Martin, 1990*a* for the alternative interpretation). Tattersall (1982) also suggests that some of the other Malagasy lemurs are more closely related to some of the Eocene adapid primates than they are to other Malagasy lemurs. In light of the adapid affinities of some of the Fayum Oligopithecinae (Simons 1990) this suggestion may not be as fanciful as it otherwise might seem. If any of these putative relationships prove true, the present day composition of the Madagascar primate faunal could be explained either by multiple events of waif dispersal between Madagascar and the mainland or as a surviving remnant of the Late Cretaceous or Early Tertiary African primate fauna shared between Madagascar and the African mainland prior to the complete separation of these two land masses. Tattersall argues in favour of this second explanation on the grounds of parsimony.

This early primate fauna would have had to have been present in the Late Cretaceous or earliest Tertiary in Africa because of the palaeogeographic history of Madagascar. Although Madagascar was separated from Africa by at least an intermittent sea barrier from the Middle Jurassic onwards (Lillegraven *et al.* 1979), major periods of low sea-level occurred in the Cenomanian (about 92 my ago) and in the Maestrichian (about 65 my ago) that might well have produced either a land bridge or closely approximated islands (Tattesall 1982). These are the most probable times during either the Late Cretaceous or the Early Cenozoic for relatively easy mammalian dispersal between Madagascar and the African mainland.

If Tattersall's interpretation is correct, it means that primates of modern aspect must have been present in Africa at the same time, or possibly before, *Purgatorius* was present in North America. As a corollary of this Africa could also have been a probable geographical area for the initial evolution of the euprimates. Even if Tattersall's interpretations are not correct, the presence of lemurs on the island of Madagascar can most easily be accounted for if euprimates were present in Africa early in the Cenozoic, before the water barrier between the island and the mainland became considerable. The absence of anthropoid primates on Madagascar would also suggest that the island was colonized by lemurs before the appearance of anthropoid primates in Africa.

Fossil evidence supporting the early existence of euprimates in Africa before their appearance at the Fayum includes *Altialtasius koulchii* from the Palaeocene of Morocco (Sigé *et al.* 1990) as well as other early primates that have been reported from Algeria (de Bonis *et al.* 1988, Godinot and Mahboubi 1992) and Tunisia (Gingerich 1990, Hartenberger and Marandat (*in press*)).

Furthermore, Martin (1990*a*) points out that the separation between the lemurs and the lorises must have occurred during the Eocene at the latest. Apart from the single loris tooth from the Fayum, the earliest well-known fossil lorisids are found in East Africa in the Early Miocene period, about 20 million years ago (Walker 1974). Based on the work of Romero-Herrera and his co-workers (1973) and on the average rate of evolution of myoglobin, he suggests that the common ancestor of the lorisids and the Malagasy lemurs must be at least twice this age. On this basis he argues for the presence of primates of modern aspect in Africa in the Eocene at the latest, and possibly much earlier.

THE POSSIBLE PRESENCE OF PRIMATES IN THE SOUTHERN CONTINENTS PRIOR TO THE COMPLETE SEPARATION OF THE CONTINENTS IN THE CRETACEOUS

The idea that euprimates might have been present in Africa during the earlier parts of the Cenozoic opens up the possibility that they were in the southern hemisphere prior to the complete separation of Africa and South America. The origin of the New World primates could then be accounted for by the complete separation of Africa and South America and the resulting genetic isolation from the Old World primates.

This argument for the origin of the New World primates is associated most strongly with the work of Hershkovitz (1972, 1977) and has come to be known as the Vicariance Origin Model (Ciochon and Chiarelli 1980). Advocates of vicariance biogeography, with which this model is consistent, argue that the main factor in speciation is the continuing creation of barriers separating and isolating ancestral breeding populations (Croizat et al. 1974). The creation of such barriers is known as a vicariance event. They also argue that the geographical distribution of an ancestral population can best be estimated by adding the distributions of the descendant populations, both fossil and extant (Nelson 1974).

Applied to the anthropoid primates, Hershkovitz has argued that the distribution of the ancestral anthropoid primate population would have been on the combined African–South American continent and that the division of the platyrrhines and catarrhines would have been initiated by the separation of the continents.

The main objection to this interpretation is that it implies a very early evolution of not only euprimates but also of the ancestral anthropoid primate stock. Current palaeogeographical analyses place the final separation of Africa from South America in the Middle Cretaceous, about 100 million years ago (Lillegraven et al. 1979; Tarling 1980). Ancestral anthropoid primates would have had to have been present in the combined southern continents at least 35 million years before the beginning of the Cenozoic, and 45 million years before the appearance of the euprimates in North America and Europe. Based on our present understanding of the patterns of mammalian evolution, this is unlikely. However, considering the relatively poor nature of the available fossil record it might (just) be in the realm of possibility (Ciochon and Chiarelli 1980).

Given the problematic nature of the primate fossil record, the best evidence in support of a vicariant origin for the New World monkeys would be similar distribution patterns for large parts of the fauna in South America and Africa (Croizat et al. 1974). Conversely, the best evidence against a vicariant explanation would be the lack of a similar distribution pattern for large parts of the fauna. It is on this point that the vicariance model can be most severely criticized on present available evidence.

Both the extant and fossil faunas of Africa and South America are very distinct (Keast 1972). The Palaeocene and Eocene mammalian fauna of Africa is dominated by endemic orders such as Hyracoidea, Embrithopoda, and Proboscidea while the contemporary fauna of South America is dominated by Xenarthra, condylarths, noto-ungulates, and marsupials (Keast 1972; Simpson 1978; Savage and Russell 1983). The only mammalian similarity, other than primates and rodents, rests with the didelphid marsupials from the Fayum of Egypt (Brown and Simons 1984). However, these fossils show close similarity to European marsupials belonging to the genus *Peratherium* and, in the view of Brown and Simons (1984), most probably represent Late Eocene immigrants to Africa rather than representatives of a larger marsupial fauna that had been present on the combined South American–African continent prior to its separation (see also Krause and Maas 1990).

Non-mammalian faunal similarity between the two continents rests with invertebrates, amphibians, reptiles, fresh-water fish, and birds (Keast 1972). These elements of the fauna can all be expected to have had a Lower Cretaceous origin (before the complete separation of the two southern continents).

Primates are also entirely absent from Australia and anthropoid primates are absent from Madagascar. Because Africa was the first of the southern continents to separate from Gondwanaland, it might be expected that if anthropoid primates, or their ancestors, were present on the combined South American and African continent they would also be present in Australia and/or Madagascar.

A final point against a great antiquity of anthropoid primates in South America is the lack of diversity of the earliest known South American fossil primates. In fact these earliest known fossil primates are represented by a single species, *Branisella boliviana* from the latest Oligocene Salla Beds of Bolivia dating back about 26 million years (MacFadden 1990). Even given the fact that there are large gaps in the fossil record of South America and that small forest-dwelling species are often poorly preserved in fossil faunas, the contrast in diversity with the Oligocene primate fauna of Africa is impressive and could be interpreted as indicating a relatively late arrival of New World monkeys in South America.

In summary, the argument that primates occupied the southern continents prior to their complete separation in the Cretaceous has little, if any, support, from known evidence. The main drawback to this argument is that only the primates (and rodents) of the southern continental mammalian fauna share a similar African and South American range. The absence of a similar range for large parts of the extant, or extinct, southern continental mammalian fauna argues against a vicariant explanation for the origin of the primates (and/or rodents).

THE RELATIVE MERITS OF THE NORTHERN CONTINENTS AND SOUTHERN CONTINENTS ORIGIN MODELS

The probability of trans-Atlantic dispersal

Although it is unlikely that ancestral anthropoid primates were present in the united southern continents before their complete separation in the Cretaceous it does not necessarily follow that it is equally unlikely that the New World monkeys had an African origin. Such an African origin any time after the Cretaceous would, however, require waif dispersal across a greater or lesser area of open water.

This interpretation of the origin of the New World primates is most closely associated with the work of Hoffstetter (Fig. 8.2) (1972, 1974, 1980) and more recently with that of Fleagle (Fleagle and Kay 1987; Fleagle, 1988). Its viability depends on the possibility that primates could have dispersed from Africa to South America across the expanding South Atlantic ocean and that the known fossil primates from the African Oligocene are the best candidates for close affinity with the New World primates.

It has been the general opinion of palaeontologists that the probability of such waif dispersal from Africa to South America would decrease with time. This is based on the fact that the half-spreading rate of the South Atlantic is about 2 cm per year (MacFadden 1990). Every 10 million years or so there would be an extra few hundred kilometres for the primates to negotiate. However, it should also be remembered that the distance across the South Atlantic need not have been open water. During the earlier parts of the Cenozoic there were areas of shallow water connecting West Africa to the northern coast of South America (the Ceara Rise and the Sierra Leone Rise) and south-west Africa with the southern coast of South America (Rio Grande Rise and Wavis Ridge). During periods of low sea-level these areas, together with the continental shelves and parts of the mid-Atlantic rise, may well have been dry land. Dispersal from one island to another would have been helped by the prevailing currents which have been reconstucted to have favoured an east to west crossing from Africa to South America (Tarling 1980).

Recent analysis of sea-level changes suggest that the first significant lowering of the sea-level in the Cenozoic occurred during the Middle Oligocene . This period of low sea-level predates the first known New World primate (*Branisella*), with a age of about 26 million years. It also postdates the earliest known African anthropoids from the Fayum which exceed 31 million years in age. The probability of waif dispersal across the South Atlantic need not have, therefore, decreased through the Cenozoic, but could have even been slightly more probable in the Middle Oligocene despite the greater distance that needed to be transversed. Such a late entry of primates into south America would be fully consistent with the recorded low diversity of Oligocene South American primates.

The parapithecids as platyrrhine ancestors

There is much agreement in the literature that the parapithecids are similar to the platyrrhine monkeys, in at least some aspects of their morphology (Simons 1972; Fleagle *et al.* 1975; Conroy 1976; Fleagle 1978; Fleagle and Kay 1987; among others). However, there is little if any agreement with Hoffstetter who argues that the parapithecids share a more recent common ancestor with the platyrrhines than they do with any other known anthropoids and should therefore, be classified with them (Szalay and Delson 1979; Kay 1980; Delson and Rosenberger 1980; Fleagle and Kay 1987).

The morphological features cited by Hoffstetter (1980), as shared in common by the South American monkeys and the parapithecids, are the presence of three premolars and a ring-shaped ectotympanic together with a series of dental features. Working on the principles of cladistic methodology, all Hoffstetter's critics emphasize the point that these resemblances between the

parapithecids and the New World monkeys are primitive (symplesiomorphic) features retained from the last common ancestor of the anthropoids and, therefore, do not necessarily indicate any special relationship between the parapithecids and the New World monkeys.

Alternative interpretations of the evolutionary position of the parapithecids place them in close phylogenetic relationship with the cercopithecid monkeys (Simons 1972, 1986; Gebo and Simons 1987) or as the sister group of all Old World Primates (Delson and Andrews 1975; Delson 1977; Szalay and Delson 1979). However, Fleagle and Kay (1987) argue that the features, upon which these relationships are based, are largely primitive or are of little import in determining evolutionary relationships because of their documented parallel occurrence in other groups of primates (see also Delson 1975; Szalay and Delson 1979; Harrison 1987). Fleagle and Kay (1987) view the parapithecids as the sister group of all other anthropoid primates including the New World monkeys (see also Harrison 1987). This interpretation positions the parapithecids on or near the stem leading to both the New World primates and the remaining primates of the Old World. This interpretation is consistent with Hoffstetter in so far as it gives the parapithecids a possible ancestral role in platyrrhine evolution. Fleagle and Kay (1987) also note that in many features the parapithecids resemble prosimians rather than anthropoid primates. These authors emphasize that these primitive features, together with the absence of a large number of features shared between later occurring New and Old World primates, offers additional evidence that the parapithecids preceded divergence between these later primates.

The possibility that the parapithecids played a role in the origin of the New World monkeys has also been made more probable in recent years by the significant redating not only of the parapithecids in Africa but also of the first known New World primate, *Branisella*. In recent years the Fayum Jebel Qatrani formation has been pushed back in time at least 5 million years. In the older literature the basalt flow overlying the formation was considered to be 25–27 my old (Simons and Wood 1968). More modern K–Ar determinations place the age of this overlying basalt flow at about 31 my old (Fleagle *et al.* 1986), while re-evaluation of the stratigraphic boundaries suggests that the entire formation should be of an older Late Eocene age (van Couvering and Harris 1981). In South America the dates for the Deseadan Beds at Salla, Bolivia, from which the *Branisella* remains have been recovered, have been revised upwards by about ten million years. Prior to recent years these beds were dated to about 33–35 my ago on the basis of faunal correlations (Marshall *et al.* 1983). New radioisotopic and magnetostratigraphic determinations support a date of about 26 my ago (MacFadden 1990). Rather than the known record of the New World monkeys (*Branisella*) predating the first African anthropoids, these African primates now predate *Branisella* by a considerable amount of time. It is also worth noting again that the very significant lowering of the sea-level in the Middle Oligocene occurs after the appearance of the African anthropoids, but before the appearance of *Branisella* in South America.

Eocene anthropoid primates in eastern Asia

Although the morphology and dating of the parapithecids offers some support for the origin of the New World primates in Africa, an additional objection to this hypothesis is the possible existence of earlier anthropoid primates in the northern hemisphere. Such primates would weaken the hypothesis that the anthropoid primates were an exclusively southern continents phenomenon and at the same time offer another, non-African, area of possible origin for the New World monkeys.

The existence of such non-African anthropoid fossils rests on the interpretation of a group of East Asian fossils of which *Pondaungia cotteri* from the Late Eocene Pondaung Formation of Burma (about 40 my ago) is arguably the most important. *Pondaungia* is known only from fragmentary jaws and teeth and has been variously interpreted as both a primate and a non-primate (von Koeingswald 1965). However, the majority of authors consider *Pondaungia* to be a primate and some consider it to be the first representative of the anthropoid primates (Maw *et al.* 1979). The main anthropoid features of *Pondaungia* are the great depth of its jaw in relation to molar crown height, its relatively thick mandibular corpus, and certain features of its postcanine dentition. Critics of the anthropoid status of this species point out that many of the sub-fossil Malagasy lemurs are characterzed by a similar jaw morphology, and that these features in *Pondaungia* could rather represent a convergence of an adapid or omomyid stock towards the anthropoid condition. They also argue that the dental similarities are found in certain Malagasy lemurs and in some New World anthropoids and therefore, are not robust features upon which to base specific evolutionary hypotheses.

Two other Asian fossils have also played a role in the argument for the Asian origin of the anthropoid primates, *Amphipithecus mogaungensis* (Pondaung Formation, Late Eocene, Burma) and *Hoanghonius stehlini* (Yuanchu, Northern China, Middle or Late Eocene). As with *Pondaungia*, both of these fossils are known only from fragmentary gnathic remains and the interpretation of these remains centres around the problem of parallelism. Briefly summarized, the mandibular corpus of *Amphipithecus* is similar in robusticity to *Pondaungia*, and both Simons (1971) and Maw *et al.* (1979) accept this as an indication of anthropoid status on the same grounds as used to interpret *Pondaungia*. However, Szalay and Delson (1979) dispute the dental affinities between this species and the Oligocene forms and, on the basis of molar form and symphyseal construction, suggest that *Amphipithecus* is most closely related to primitive adapids. They suggest that any similarity between *Amphipithecus* and the Oligocene Fayum primates, particularly in premolar morphology, results from parallel development.

The importance of *Hoanghonius* lies in the similarity between its lower M2 and that of

Oligopithecus. Szalay and Delson (1979) suggest that, based on the similarity in the formation of the talonid in these two fossils, it is possible that if *Oligopithecus* is a catarrhine *Hoanghonius* is also a catarrhine (and by inference an anthropoid). However, similarity in a single dental feature of unknown robusticity is, at the best, weak evidence upon which to accept anthropoid status for this fossil. Szalay and Delson (1979) recognize this point by classifying *Hoanghonius* in the Family Omomyidae, *incertae sedis*

Acceptance of any of these fossils as anthropoid primates is largely a matter of taste. However, their anthropoid status is central to the two hypotheses for the origin of the New World primates that argue against a higher primate presence in Africa in the earlier Cenozoic. In particular, version three of the Southern Continents Model argues that they represent the origin of the anthropoids in eastern Asia and that these primates dispersed into Africa sometime in the Late Eocene (Ciochon and Chiarelli 1980). To some, their anthropoid status is even supported, in a circular fashion, by the geographic position halfway between Africa and the Americas (Maw *et al.* 1979) and their potential importance to the Northern Continents Model for the origin of the New World primates. Gingerich (1980) suggests that they represent the origin of the anthropoid primates in Asia from an adapid ancestor According to this model the ancestors of the platyrrhines would have migrated from Asia across the Bering Strait land bridge and then would have proceeded through North America and entered South America by rafting across the Caribbean by the latest Eocene. Alternatively, Delson and Rosenberger (1980) propose that the anthropoid primates arose in North America from an omomyid ancestor in the Late Eocene. The ancestors of the New World monkeys would have migrated from North America to South America by rafting across the Caribbean in the latest Eocene, while the ancestors of the catarrhines would have migrated across the Bering Strait to Asia (represented by *Pondaungia*, *Amphipithecus*, and *Hoanghonius*) and would

have entered Africa by rafting across the Tethys Sea in the latest Eocene or earliest Oligocene.

In relation to the origin of the New World monkeys the main objection to both of these hypotheses is the total absence of anthropoid primate fossils in the Eocene of North America (Fleagle and Kay 1987). The North American primate fauna is relatively well known and it would seem improbable that if the anthropoids existed there during the Eocene that they have remained totally unrecognized.

There is an additional problem with the entry of the ancestral platyrrhines into South America and also with the entry of the ancestral African primates into that continent during the Eocene. In the Americas the reconstructed ocean currents between North and South America would have moved from east to west in the ancestral Caribbean and not from north to south (Tarling 1980). They, therefore, would not facilitate a Northern Continents origin for the New World primates.

In the Old World, the main problem with the dispersal of primates from Eurasia to Africa is that any connection across the Tethys sea in the Eocene most probably was a filter connection. This idea of a filter connection is supported by the conspicuous absence in the Fayum fauna of northern continental mammals such as the fissipid carnivores, perissodactyls, heridomyid, and cricetid rodents and palaertic artiodactyls (Savage and Russell 1983). Those elements of the Fayum fauna in common with Eurasia are the hyaenodontids, cebochoerieds, anthracotherids, embrithopoids (Savage and Russell 1983), the marsupials (Brown and Simons 1984), and the proboscidians and cetaceans (West 1980).

Considering the occurrence of at least one euprimate in the Palaeocene of Morocco, as well as the Early to Middle Eocene age of the earliest known African anthropoid, and also the great diversity of the Fayum primate fauna it would seem unnecessary to postulate that evolution would proceed rapidly enough or that sufficient species would negotiate the filter connection to result in the known Late Eocene/Early Oligocene primate fauna.

In view of this evidence, if the East Asian fossils are actually anthropoid primates, it is possible that their existence in eastern Asia could be explained equally well by a migration from Africa to Asia as by a migration from Asia to Africa in the Eocene (see Cooke 1968 and Gingerich and Rose 1977 for arguments based on the non-primate fauna in support of this hypothesis).

SUMMARY AND CONCLUSIONS

Of the viable hypotheses for the origin of the New World primates, the foregoing discussions have demonstrated that the Southern Continents Vicariance Model (Fig. 2a) (Hershkovitz 1972, 1977) is the least likely. At present, there is no convincing evidence that primates of any form existed in the southern continents prior to the initial formation of the South Atlantic ocean in the later part of the Cretaceous. More probable is an African origin of the New World monkeys that involves dispersal across the South Atlantic sometime prior to the Late Oligocene and the first appearance of fossil New World primates at about 26 million years ago. However, it should also be emphasized that there are insufficient grounds upon which to confidently reject the North Continents Model on the basis of any hard and unequivocal evidence. As a consequence the following hypotheses for the origin of the anthropoid primates and of the New World monkeys are currently viable.

1. The anthropoid primates had their origin in the southern continents and the platyrrhine monkeys had their origin in Africa (Fig. 2b) (Hoffstetter 1972, 1974, 1980, 1988; Fleagle and Kay 1987; Fleagle 1988).
2. The anthropoid primates had their origin in the northern continents and the platyrrhine monkeys had their origin in North America (Figs. 1a and b) (Gingerich 1980; Delson and Rosenberger 1980).
3. The anthropoid primates had their origin in the northern continents and the platyrrhine

monkeys had their origin in Africa (Fig. 2c) (Ciochon and Chiarelli 1980).

At present the first of these hypotheses, the Hoffstetter-Fleagle hypothesis, is most consistent with current evidence. The early existence of euprimates in Africa facilitates the dispersal of ancestral lemurs to Madagascar. The evolution of the anthropoid primates on that continent is consistent with the diversity of the Fayum anthropoid fauna and is also consistent with the evidence that anthropoid primates may have enjoyed a relatively long period of evolution prior to the separation of the New and Old World primates. Furthermore, the Fayum parapithecids are the best known candidates for the immediate ancestry of the New World monkeys. This hypothesis is strengthened by new evidence that euprimates occupied North Africa in the Palaeocene and Eocene and also by the fact that the Eocene euprimates of North America and Europe are part of an immigrant fauna that originated in more southern latitudes.

The Ciochon and Chiarelli (1980) hypothesis, which proposes a northern continents origin for the anthropoid primates and an African origin for the New World monkeys, is consistent with the arguable anthropoid status of *Pondaungia* in eastern Asia and with the fact that the parapithecids are the best known candidates for the immediate ancestry of the New World monkeys. Considering the marked lowering of sea-level in the Middle Oligocene it is also no longer weakened by the suggestion that a post-Eocene dispersal from Africa to South America would be any less likely than a dispersal earlier in the Cenozoic. However, this hypothesis does not easily account for the diversity of the Early Oligocene primate fauna from the Fayum. Nor does it account or the presence of the lemurs in Madagascar without speculating that the strepsirhine primates were much earlier immigrants into Africa than were the anthropoid primates.

The remaining Northern Continents Model (Delson and Rosenberger 1980; Gingerich 1980) which argues for the evolution of the anthropoid primates in the northern hemisphere and for the immediate ancestry of the New World monkeys in North America suffers from the same weaknesses as the previous hypothesis, while at the same time ignoring the potential role of the parapithecids in the ancestry of the New World monkeys. It is also weakened by the total absence of anthropoid primates in the fossil record of North America, even though the Eocene record on this continent is relatively well known. By postulating a North American origin of the New World monkeys it also ignores the evidence which suggests that the reconstructed ocean currents would have rendered waif dispersal unlikely (Tarling 1980).

Given the present state of knowledge, the Hoffstetter–Fleagle African-centred model appears to account for more of the available evidence for the distribution and evolution of fossil and extant primates than do the other viable models. This hypothesis is strengthened by the newly discovered pre-Oligocene African primate fauna, by the redating of both the Fayum and the Deseadan Beds of Bolivia and by the recognition that there was a period of markedly low sea-level during the Oligocene. However, the early Cenozoic fossil records of Asia and of the southern continents are still poorly known, and until we have a more comprehensive knowledge of primates from these periods, the area and time of origin of the platyrrhine monkeys will remain one of the most controversial topics in primate evolution.

ACKNOWLEDGEMENTS

I would like to thank Wilma George for inviting me to contribute to this volume and Professor R. Lavocat for guiding the project through to completion after Wilma's untimely death. This paper was first written in 1985. Since then it has gone through a number of drafts as new fossil material has been recovered. I am happy to say that the argument for an African origin of the New World primates is stronger now than it was at that time. I would especially like to thank M. Godinot, R. Lavocat, R. D. Martin, and P. Andrews for commenting on earlier drafts of this manuscript and/or making me aware of new discoveries even before they appeared in print.

REFERENCES

Baba, M., Darga, L., and Goodman, M. (1980). Biochemical evidence on the phylogeny of Anthropoidea. In *Evolutionary biology of the New World monkeys and continental drift* (ed. R. L. Ciochon and A. B. Chiarelli), pp. 423–33. Plenum Press, New York.

Beard, K. C. (1990). Gliding behaviour and palaeoecology of the alleged primate family Paromomyidae (Mammalia, Dermoptera). *Nature*, **345**, 340–1.

de Bonis, L., Jaeger, J. J., Coffait, B., and Coffait, P.-E. (1988). *Comptes Rendus de l'Académie des Sciences Paris*, **306**, 929–34.

Bown, T. M. and Simons, E. L. (1984). First record of marsupials (Metatheria: Polyprotodonta) from the Oligocene in Africa. *Nature*, **308**, 447–9.

Cachel, S. M. (1979). A functional analysis of the primate masticatory system and the origin of the anthropoid post-orbital septum. *American Journal of Physical Anthropology*, **50**, 1–18.

Cartmill M. (1975). Strepsirhine basicranial structures and the affinities of the Cheirogaleidae. In *Phylogeny of the primates* (ed. W. P. Luckett and F. S. Szalay), pp. 313–54. Plenum Press, New York.

Ciochon, R. L. and Chiarelli, A. B. (1980). Paleobiogeographic perspectives on the origin of the Platyrrhini. In *Evolutionary biology of the New World monkeys and continental drift* (ed. R. L. Ciochon and A. B. Chiarelli), pp. 459–93. Plenum Press, New York.

Conroy, G. C. (1976). Primate postcranial remains from the Oligocene of Egypt. *Contributions to Primatology*, **8**, 1–134.

Conroy, G. C. (1980). Ontogeny, auditory structures and primate evolution. *American Journal of Physical Anthropology*, **52**, 443–51.

Cooke, H. B. S. (1968). Evolution of mammals on the southern continents, Part II: The fossil mammals of Africa. *The Quarterly Review of Biology*, **43**, 234–64.

Cooke, H. B. S. (1972). The fossil mammals of Africa. In *Evolution, mammals and southern continents* (ed. A. Keast, F. C. Erk, and B. Glass), pp. 89–139. State University of New York Press, Albany.

Couvering, J. A. van and Harris, J. A. (1991) Late

Eocene age of Fayum mammal faunas. *Journal of Human Evolution*, **21**, 241–60

Cracraft, J. (1974). Continental drift and vertebrate distribution. *Annual Review of Ecology and Systematics*, **5**, 215–61.

Croizat, L., Nelson, G., and Rosen, D. E. (1974). Centers of origin and related concepts *Systematic Zoology*, **23**, 265–87.

Dashzeveg, D. and McKenna, M. C. (1977). Tarsoid primate from the early Eocene of the Mongolian Peoples Republic. *Acta Palaeontological Polonica*, **22**, 119–37.

Delson, E. (1975). Toward the origin of the Old World monkeys. Evolution des Vertébrés – Problèmes Actuels de Paléontologie. *Actes CNRS Colloques Internationales*, **218**, 39–50.

Delson, E. (1977). Catarrhine phylogeny and classification: principles, methods and comments. *Journal of Human Evolution*, **6**, 433–59.

Delson, E. and Andrews, P. (1975). Evolution and interrelationships of the catarrhine primates. In *Phylogeny of the primates: a multidisciplinary approach* (ed. W. P. Luckett and F. S. Szalay), pp. 405–46. Plenum Press, New York.

Delson, E. and Rosenberger, A. L. (1980). Phyletic perspectives on platyrrhine origins and anthropoid relationship. In *Evolutionary biology of the New World monkeys and continental drift* (ed. R. L. Ciochon and A. B. Chiarelli), pp. 445–58. Plenum Press, New York.

Fleagle, J. G. (1978). Size distributions of living and fossil primate faunas. *Paleobiology*, **4**, 67–76.

Fleagle, J. G. (1988). *Primate adaptation and evolution.* Academic Press, San Diego.

Fleagle, J. G. and Kay R. F. (1987). The phyletic position of the Parapithecidae. *Journal of Human Evolution*, **16**, 485–532.

Fleagle, J. G., Simons, E. L., and Conroy, G. C. (1975). Ape limb bone from the Oligocene of Egypt. *Science*, **189**, 35–7.

Fleagle, J. G., Bown, T. M., Obradovich, J. D., and Simons, E. L. (1986). Age of the earliest African anthropoids. *Science*, **234**, 1247–9.

Gazin, C. L. (1958). A review of the Middle and Upper Eocene Primates of North America. *Smithsonian Miscellaneous Collection*, **136**, 1–112.

Gebo, D. L. and Simons, E. L. (1987). Morphology and locomotor adaptations of the foot in early Oligocene anthropoids. *American Journal of Physical Anthropology*, **74**, 83–101.

Gingerich, P. D. (1977). Radiation of Eocene Adapidae in Europe. *Geobos, Mem. Spec.*, **1**, 165–82.

Gingerich, P. D. (1979). *Indraloris* and *Sivaladapis*: Miocene adapid primates from the Siwaliks of India and Pakistan. *Nature*, **279**, 415–16.

Gingerich, P. D. (1980). Eocene Adapidae, paleobiogeography, and the origin of South American Platyrrhini. In *Evolutionary biology of the New World monkeys and continental drift* (ed. R. L. Ciochon and A. B. Chiarelli), pp. 123–38. Plenum Press, New York.

Gingerich, P. D. (1981). Early Cenozoic Omomyidae and the evolutionary history of tarsiiform primates. *Journal of Human Evolution*, **10**, 345–74.

Gingerich, P. D. (1986). Early Eocene *Cantius torresi*— oldest primate of modern aspect from North America. *Nature*, **319**, 319–21.

Gingerich, P. D. (1990). African dawn for primates. *Nature*, **346**, 411.

Gingerich, P. D. and Rose, K. D. (1977). Preliminary report on the American Clark Fork mammal fauna and its correlation with similar faunas in Europe and Asia. *Geobios. Mem. special*, **1**, 39–45.

Gingerich, P. D. and Sahni, A. (1984). Dentition of *Sivaladapis nagrii* (Adapidae) from the late Miocene of India. *International Journal of Primatology*, **5**, 63–79.

Godinot, M. and Mahboubi, M. (1992). Earliest known simian primate found in Algeria. *Nature*, **357**, 324–6.

Gregory, W. K. (1920). Evolution of the human dentition, Part IV: The South American monkeys, *Journal of Dental Research*, **2**, 404–26.

Gregory, W. K. (1921). On the structure and relations of *Northarctus*, an American Eocene primate. *Memoirs of the American Museum of Natural History*, **3**, 49–243.

Hartenberger, J.-L. and Marandat, B. A new genus and species of an early Eocene Primate from North Africa. *Journal of Human Evolution (in press)*.

Harrison, T. (1987). The phylogenetic relationships of the early catarrhine primates: a review of the current evidence. *Journal of Human Evolution*, **16**, 41–80.

Hershkovitz, P. (1972). The recent mammals of the Neotropical region: A zoogeographic and ecological review. In *Evolution, mammals and southern continents* (ed. A. Keast, F. C. Erk, and B. Glass), pp. 311–431. State University of New York Press, Albany.

Hershkovitz, P. (1977). *Living New World monkeys*

(Platyrrhini), with an introduction to primates, Vol. 1. University of Chicago Press, Chicago.

Hoffstetter, R. (1972). Relationships, origins, and history of the ceboid monkeys and caviomorph rodents: a modern reintepretation. *Evolutionary Biology*, **6**, 323–7.

Hoffstetter, R. (1974). Phylogeny and geographical deployment of the primates. *Journal of Human Evolution*, **3**, 327–50.

Hoffstetter, R. (1980). Origin and deployment of New World monkeys emphasizing the southern continents route. In *Evolutionary biology of the New World monkeys and continental drift* (ed. R. L. Ciochon and A. B. Chiarelli) pp. 103–22. Plenum Press, New York.

Hoffstetter, R. (1988). Origine et évolution des primates non-humains du nouveau monde. In *L'évolution dans sa réalité et ses diverses modalités* (M. Maurois, Org.), pp. 133–70. Masson and Singer-Polignac, Paris.

Kay, R. F. (1980). Platyrrhine origins: a reappraisal of the dental evidence. In *Evolutionary biology of the New World monkeys and continental drift* (ed. R. L. Ciochon and A. B. Chiarelli), pp. 159–87. Plenum Press, New York.

Kauy, R. F., Thorington, R. W., and Houde, P. (1990). Eocene plesiadapiform shows affinities with flying lemurs, not primates. *Nature*, **345**, 324–44.

Keast, A. (1972). Comparisons of contemporary mammal faunas of the southern continents. In *Evolution, mammals and southern continents* (ed. A. Keast, F. C. Erk, and B. Glass), pp. 433–501. State University of New York Press, Albany.

Koenigswald, G. H. R. von (1965). Critical observations upon the so-called higher primates from the upper Eocene of Burma. *Proc. Kon. Nederlandse Akad. Wetensch.*, Ser. B. **68**, 165–7.

Krause, D. W. and Maas, M. C. (1990). The biogeographic origins of late Palaeocene–early Eocene mammalian immigrants to the Western Interior of North America. In *Dawn of the Age of Mammals in the northern part of the Rocky Mountains interior, North America* (ed. T. M. Brown and K. D. Rose), pp. 71–105. Geological Society of America, Special Paper 243.

Lavocat, R. (1974). The interrelationships between the African and South American rodents and their bearing on the problem of the origin of South American monkeys. *Journal of Human Evolution*, **3**, 323–6.

Lavocat, R. (1977). Sur l'origine des faunes sud-americaines de Mammifères du Mésozoique terminal et du Cenozoique Ancien. *Comptes Rendus de l'Académie des Sciences Paris* Sér. D, **285**, 423–6.

Lavocat, R. (1980). The implications of rodent palaeontology and biogeography to the geographical sources and origin of platyrrhine primates. In *Evolutionary biology of the New World monkeys and continental drift* (ed. R. L. Ciochon and A. B. Chiarelli), pp. 93–102. Plenum Press, New York.

Lillegraven, J. A., Kielan-Jaworowska, Z., and Clemens, W. A. (ed.) (1979). *Mesozoic mammals*. University of California Press, Berkeley.

Loomis, F. B. (1911). The adaptation of the primates. *American Naturalist*, **45**, 479–92.

Luckett, W. P. (1980). Monophyletic or diphyletic origins of Anthropoidea and Hystricognathi: Evidence of the fetal membranes. In *Evolutionary biology of the New World monkeys and continental drift* (ed. R. L. Ciochon and A. B. Chiarelli), pp. 347–68. Plenum Press, New York.

MacFadden, B. J. (1990). Chronology of Cenozoic primate localities in South America. *Journal of Human Evolution*, **19**, 7–22.

McKenna, M. C. (1980). Eocene paleolatitude, climate and mammals of Ellesmere Island. *Palaeogeography, Paleoclimatology, Palaeocology*, **30**, 349–62.

McKenna, M. C. (1983). Cenozoic paleogeography of North Atlantic land bridges. In *Structure and development of the Greenland Scotland Ridge* (ed. Bott, M. H. P., Saxov, S., Talwanj, M., and Thiede, J.), pp. 351–99. Plenum Press, New York.

Marshall, L. G., Hoffstetter, G. R., and Pascual, R. (1983). Mammals and stratigraphy of the continental mammal-bearing Tertiary of South America. *Palaeovertebrata, Mem. Extraord.*, 1–93.

Martin, R. D. (1990*a*). *Primate origins and evolution.* Chapman and Hall, London.

Martin, R. D. (1990*b*). Some relatives take a dive. *Nature*, **345**, 291–2.

Maw, B., Ciochon, R. L., and Savage, D. E. (1979). Late Eocene of Burma yields earliest anthropoid primate, *Pondaungia cotteri. Nature*, **282**, 65–7.

Nelson, G. (1974). Historical biogeography: an alternative formalization. *Systematic Zoology*, **23**, 555–8.

Pan, Y. and Wu, R. (1986). A new species of *Sinoadapis* from the hominoid site, Lufeng. *Acta Anthropologica Sinica*, **5**, 39–50.

Pickford, M. (1986). Première découverte d'une faune mammalienne terrestre paléogène d'Afrique subsaharienne. *Comptes rendus des séances de l'Académie des Sciences, Series II*, **302**, 1205–10.

Romero-Herrera, A. E., Lehmann, H., Joysey, K. A., and Friday, A. E. (1973). Molecular evolution of myoglobin and the fossil record: a phylogenetic synthesis. *Nature*, **246**, 389–95.

Rose, K. D. (1980). Clarkforkian land-mammal age: revised definition zonation and tentative intercontinental correlations. *Science*, **208**, 744–6.

Rose, K. D. and Krause, D. W. (1984). Affinities of the primate *Altanius* from the early Tertiary of Mongolia. *Journal of Mammalogy*, **65**, 721–6.

Sarich, V. M. (1970). Primate systematics with special reference to Old World monkeys. In *Old World monkeys* (ed. J. R. Napier and P. H. Napier), pp. 175–226. Academic Press, New York.

Sarich, V. M. and Cronin, J. E. (1980). South American mammal molecular systematics, evolutionary clocks and continental drift. In *Evolutionary biology of the New World monkeys and Continental drift* (ed. R. L. Ciochon and A. B. Chiarelli), pp. 399–421. Plenum Press, New York.

Sigé, B., Jaeger, J.-J., Sudre, J., and Vianey-Liaud, M. (1990). *Altiatlasius koulchii* n.gen. et sp., primate omomyidé du Paléocène supérieur du Maroc, et les origines des euprimates. *Palaeontographica*, A, **214**, 31–56.

Simons, E. L. (1961). The dentition of *Ourayia*—its bearing on relationships of omomyid prosimians. *Postilla*, **54**, 1–233.

Simons, E. L. (1971). Relationships of *Amphipithecus* and *Oligopithecus. Nature*, **232**, 489–91.

Simons, E. L. (1972). *Primate evolution.* Macmillan, New York.

Simons, E. L. (1986). *Parapithecus grangeri* of the African Oligocene: an archaic catarrhine without lower incisors. *Journal of Human Evolution*, **15**, 205–13.

Simons, E. L. (1989). Description of two genera and species of Late Eocene Anthropoidea from Egypt. *Proceedings of the National Academy of Sciences (USA)*, **86**, 9956.

Simons, E. L. (1990). Discovery of the oldest known anthropoidean skull from the Paleogene of Egypt. *Science*, **247**, 1567–9.

Simons, E. L. and Bown, T. M. (1985). *Afrotarsius chatrathi*, first tarsiiform primate (?Tarsiidae) from Africa. *Nature*, **313**, 475–7.

Simons, E. L. and Wood, A. E. (1968). Early Cenozoic mammalian faunas, Fayum Province, Egypt. *Bulletin of the Peabody Museum of Natural History*, **28**, 1–105.

Simons, E. L., Bown, T. M., and Rasmussen, D. T. (1987). Discovery of two additional prosimian primate families (Omomyidae, Lorisidae) in the African Oligocene. *Journal of Human Evolution*, **15**, 431–7.

Simpson, G. G. (1978). Early mammals in South America: fact, controversy and mystery. *Proceedings of the American Philosophical Society*, **122**, 318–28.

Sloan, R. E. (1969). Cretaceous and Paleocene terrestrial communities of western North America. *Proceedings of the North American Paleontological Convention, Sept. 1969, Part E.* pp. 427–53.

Sudre, J. (1975). Un Prosimian du Paleogene ancien du Sahara nord occidental: *Azibus Trerki* n.g. n.sp. *Comptes Rendus de l'Académie Sciences (Paris)*, **280D**, 1539–42.

Szalay, F. S. (1975). Phylogeny, adaptations, and dispersal of Tarsiiform primates. In *Phylogeny of the primates* (ed. W. P. Luckett and F. S. Szalay), pp. 357–404. Plenum Press, New York.

Szalay, F. S. and Delson, E. (1979). *Evolutionary history of the primates.* Academic Press, New York.

Szalay, F. S. and Katz, C. C. (1973). Phylogeny of lemurs, galagos, and lorises. *Folia Primatologia*, **19**, 88–103.

Tarling, D. (1980). The geologic evolution of South America with special reference to the last 200 million years. In *Evolutionary biology of the New World monkeys and continental drift* (ed. R. L. Ciochon and A. B. Chiarelli), pp. 1–41. Plenum Press, New York.

Tattersall, I. (1982). *The primates of Madagascar,* Columbia University Press, New York.

Tattersall, I. and Schwartz, J. H. (1974). Craniodental morphology and the systematics of the Malagesy lemurs (Primates, Prosimii). *Anthropological Papers of the American Museum of Natural History*, **52**, 139–92.

Tattersall, I. and Schwartz, J. H. (1975). Relationships among the Malagasy lemurs: The craniodental evidence. In *Phylogeny of the primates* (ed. W. P. Luckett and F. S. Szalay), pp. 299–312. Plenum Press, New York.

Thomas, H., Sen, S., Roger, J., and Al-Sulaimani, Z. (1991). The discovery of *Moeripithecus markgrafi* Schlosser (Propliopithecidae, Anthropoidea, Primates), in the Ashawq Formation (Early Oligocene of Dhofar Province, Sultanate of Oman). *Journal of Human Evolution*, **20**, 33–49.

Walker, A. C. (1974). A review of the Miocene Lorisidae of East Africa. In *Prosimian biology* (ed. R. D. Martin, G., A. Doyle, and A. C. Walker), pp. 435–47. Duckworth, London.

West, R. M. (1980). Middle Eocene large mammal assemblage with Tethyan affinities, Ganda Kas region, Pakistan. *Journal of Paleontology*, **54**, 508–33.

Wible, J. R. and Covert, H. H. (1987). Primates: cladistic diagnosis and relationships. *Journal of Human Evolution*, **16**, 1–22.

Wolfe, J. A. (1975). Some aspects of plant geography of the northern hemisphere during the late Cretaceous and Tertiary. *Annals of the Missouri Botanical Garden*, **62**, 264–79.

Wolfe, J. A. (1978). A paleobotanical interpretation of Tertiary climates in the northern hemisphere. *American Scientist*, **66**, 695–703.

Wolfe, J. A. (1980). Tertiary climates and floristic relationships at high latitudes in the northern hemisphere. *Paleogeography, Paleoclimatology and Paleoecology*, **30**, 313–23.

Wu, R. and Pan, Y. (1985). A new adapid primate from the Lufeng Miocene, Yunnan. *Acta Anthropologica Sinica*, **4**, 1–6.

9 THE STRANGE RODENTS OF AFRICA AND SOUTH AMERICA

Wilma George

The name rodent—from the Latin *rodo*, to gnaw—indicates the chief characteristic of the group. Rodents have ever-growing incisors, with enamel on the outside only; and they have a jaw which can be aligned either to gnaw with the incisors or to chew with the cheek teeth. The two working positions necessitate moving the jaw fore and aft. The jaw joint is, therefore, long and loose-fitting; there is a gap, the diastema, between the two sets of teeth; the two sides of the lower jaw are only loosely connected in front; and the masseter muscles—the chewing muscles—are complex. These arrangements allow for a great variety of jaw movements. In other respects rodents have few unique characteristics.

The rodents can be divided into four main suborders according to the arrangement of the masseter muscles and the consequent four jaw shapes. The masseter muscles are in three layers. The earliest arrangement is called protrogomorphous (Wood 1965) which means early or first rodent shape (Fig. 9.1). It is well known in fossils and represented today by the North American mountain beavers (suborder Protrogomorpha, family Aplondontidae). The superficial chewing muscle—the outer masseter superficialis which draws the lower jaw forward—arises from the front of the bony arch below the orbit and inserts at the back of the lower jaw. The middle layer, the masseter lateralis, arises on the underside of the bony arch below the orbit and runs to the lower edge of the lower jaw. It pulls the jaw vertically upward. The deep layer, the masseter medialis, arises from the inner side of the arch and runs to the upper and outer part of the lower jaw. It pulls the jaw diagonally upward and holds it shut.

The suborder Sciuromorpha—with a squirrel-shape jaw—is slightly modified from the protrogomorphous jaw by the lengthening of the masseter lateralis to give better gnawing and extended vertical movement. In the suborder Myomorpha—mouse-shape jaw—the masseter medialis has increased in length, too. It runs from the front of the orbit, with a small branch through the canal that carries nerves and blood vessels through the orbit to the snout (Fig. 9.1b). It

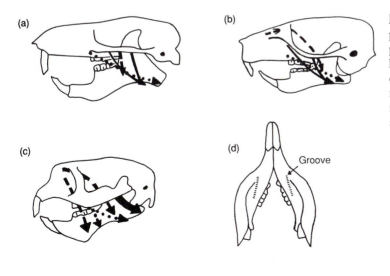

(a)

(b)

(c)

(d) Groove

Fig. 9.1 Skull shape and muscle position in (a) protrogomorphous pattern, (b) myomorphous pattern, (c) hystricomorphous pattern, (d) hystricognathous pattern. Dotted iine, masseter superficialis; solid line, masseter lateralis; dashed line, masseter medialis.

increases diagonal movement. The myomorphous jaw is very versatile (Hiiemae 1978, Kesner 1980). The fourth type of jaw is the porcupine-shape jaw of the suborder Hystricomorpha (Fig. 9.1c). The masseter lateralis is relatively unmodified from the protrogomorphous condition but the masseter superficialis wraps into a groove at the base of the anterior part of the mandible. It is the masseter medialis, however, that is the most modified. It is huge and starts far forward over the snout and passes through an enlarged canal through the front of the orbit to insert well forward on the mandible. This gives a powerful diagonal chopping and chewing action and holds the upper and lower teeth tightly together.

The shape of the lower jaw, too, is modified in most hystricomorph rodents. The curious shape of these lower jaws is called hystricognathous.

The hystricognathous jaw has the particular angle and the chewing teeth outside and to the side of the line of the incisors, giving a sharp outwardly deflected angle to the jaw behind the incisors (Fig. 9.1d). This pattern allows further lengthening of the masseter medialis and gives it an increased insertion surface; and it favours the lengthening of another muscle, the pterygoideus internus, which controls lateral movements of the jaw. This pattern also increases the power of the jaw. Unlike most mammals, some of the hystricognathous rodents can chew with both sides of the jaw at the same time (Woods 1972; Byrd 1981). With the development of a special type of enamel on the incisors, the hystricomorphous–hystricognathous jaw becomes eminently suitable for processing tough vegetation such as grass (Beecher 1979).

The hystricognathous lower jaw is considered to be of such fundamental importance that it is often used to replace Hystricomorpha as a suborder. Unfortunately not all rodents with hystricomorphous jaws are also hystricognathous, nor are all hystricognaths hystricomorphous. This leads to problems with the classification.

In addition to jaw muscles, the hystricognath–hystricomorph rodents have a few other skeletal and muscular characteristics in common as well as some important reproductive features which distinguish them from other rodents.

To biogeographers, the hystricognath–hystricomorph rodents are a puzzle. There are hystricognath–hystricomorphs in South America (often called caviomorphs): guinea-pigs, chinchillas, porcupines, for example; and there are hystricognath–hystricomorphs in Africa: cane rats, rock rats, mole rats, and porcupines, for example. Table 9.1 lists the modern hystricognath–hystricomorph families. Apart from South America and Africa there are immigrant New World porcupines in North America; and there are Old World porcupines in Asia and Europe. How can this distribution be explained?

Equally puzzling to the biogeographers is the distribution of hystricognath–hystricomorph fossils. The rocks of the lower Oligocene of 32–36 million[1] ago indicate a simultaneous radiation (in geological terms) of these rodents in South America and Africa. Both continents were more or less isolated from the rest of the world's land masses at that time. How did the ancient hystricognath–hystricomorph rodents get there?

CLASSIFICATIONS OF HYSTRICOGNATH–HYSTRICOMORPH RODENTS

The classification of the hystricognath–hystricomorph rodents is far from established, Table 9.2.

The first explicit use of the masseter muscles as a basis of rodent classification was that of Brandt in 1855. He combined all the South American families with Asian and African hystricid porcupines, African thryonomyid cane rats, and ctenodactylid gundis into a suborder Hystricomorpha. In 1896, Thomas emphasized the fusion of the tibia and fibula bones of the hind leg and brought

Macfadden *et al.* (1985) propose shifting these Deseadan rocks of 10 million years upwards (Upper Oligocene–Lower Miocene). The various arguments for entering the dating must be examined with prudence. Pascual (in lit.) disagrees with MacFadden's dating (R.L.).

Table 9.1 Modern hystricognath–hystricomorph rodent families

South America		
Hydrochoeridae: capybaras	} Sciurognath	
Caviidae: guinea-pigs, maras		
Dasyproctidae: agoutis, pacas		} Hystricomorph
Dinomyidae: pacarans		'Caviomorphs'
Chinchillidae: chinchillas, viscachas		
Echimyidae: spiny rats, casiraguas		
Ctenomyidae: tuco-tucos	} Hystricognath	
Octodontidae: degus		
Capromyidae: hutias, coypu		
Abrocomyidae: rat chinchillas		
Erethizontidae: New World porcupines		
North America		
Erethizontidae: late immigrant from the south		
Africa		
Pedetidae: spring hares	} Sciurognath	
Anomaluridae: scale-tail flying squirrels		
Ctenodactylidae: gundis		
Thryonomyidae: cane rats		} Hystricomorph
Petromuridae: African rock rats		
Hystricidae: Old World porcupines	} Hystricognath	
Bathyergidae: blesmols or mole rats		Protrogomorph
Asia and Europe		
Hystricidae		

in the pedetid spring hares. Tullberg using an array of characters, but relying heavily on the form of the lower jaw, made the group Hystricognathi in 1899. The Hystricognathi brought in the African bathyergid mole rats, but excluded the ctenodactylids which are hystricomorphous but not hystricognathous. Tullberg separated American porcupines and Old World porcupines into two families for the first time.

This century, Ellerman (1940), studying skulls, brought African thryonomyids, petromurid rock rats, and hystricids into a superfamily with the majority of South American hystricognaths, but he distanced the caviid guinea-pigs and hydrochoerid capybaras from this superfamily because the lower jaws are only weakly hystricognathous. He distanced, but still included, bathyergid mole rats in the main group but removed ctenodactylids, anomalurids, and pedetids.

In 1955, Wood, using an extensive knowledge of fossil jaws and teeth, made the suborder Caviomorpha for the South American families and separated them from the Hystricomorpha (hystricids, thryonomyids, and petromurids) and the Bathyergomorpha (bathyergids only). Later (1965) Wood put hystricids into a separate Hystricomorpha and united the other African and the American groups into the Hystricognathi. Still the ctenodactylids, pedetids, and anomalurids were nowhere.

Meanwhile Landry (1957) had surveyed eight anatomical characters—which included soft parts as well as skulls—and brought back ctenodactylids.

Table 9.2 Classification of the hystricognath–hystricomorph rodents +, included in the main classification; −, not included

	South American families	Erethizontids	Hystricids	Thryonomyids Petromurids	Bathyergids	Ctenodactylids	Pedetids	Anomalurids
Brandt 1855	Hystricomorpha				−	+	−	−
Thomas 1896	Hystricomorpha				−	+	+	−
Tullberg 1899	Hystricognathi					−	−	−
Wood 1955	Hystricognathi					−	−	−
	Caviomorpha		hystricomorpha		Bathyer-gomor-pha			
Wood 1965, 1982	Caviomorpha		Hystrico-morpha	Phio-morpha	Bathyer-gomor-pha	−	−	−
Landry 1957	Hystricomorpha					+	−	−
Lavocat 1973	Hystricognathi					−	?	?
	Caviomorpha		Phibmorpha			−	Anomaluromorpha	
Chaline and Mein 1979	Hystricognathi					−	−	−
Woods 1972	Hystricognathi		−	Hystricognathi		−	−	−
Bugge 1974	Cavio-morpha	Ereth-izonto-morpha	Hystricomorpha			−	−	−

In 1973 Lavocat—again basing the classification on an extensive knowledge of fossil anatomy—used the suborder Phiomorpha for thryonomyids and petromurids (with which Wood more or less agrees) but included Bathyergids (with which Wood does not agree). Separate, but under the general heading of Hystricognathi, were pedetids and anomalurids. Ctenodactylids had gone to the squirrel group.

Closely following Lavocat, Chaline and Mein (1979) grouped hystricids, thryonomyids, petromurids, and bathyergids, together but, for a change, put pedetids and ctenodactylids together, though not as hystricomorphs.

A quite new analysis, using the form of the cranial arteries, also brought Bugge (1974) to keep hystricids close to thryonomyids, petromurids, and bathyergids. But Bugge separated New World porcupine erethizontids into a suborder of their own as he did with anomalurids (to which were now joined pedetids).

From a detailed study of head and shoulder musculature, Woods (1972) distanced hysticids once again and, in a later study (Woods and Hermanson 1985), concluded that there was some justification for Bugge to separate erethizontids from other South American hystricognaths.

A study of the thoracic aorta, the vena cava, the azygos, and the gonadal veins led George (1981) also to distance the erethizontids. Anomalurids and pedetids seemed to have little in common with other hystricognaths but the bathyergids and ctenodactylids were closer.

In 1985, Luckett, on the basis of tooth development and characteristics of the embryonic membranes, included bathyergids in the Hystricognathi and concluded that ctenodactylids were a sister group to all the Hystricognathi. Anomalurids and pedetids remained as anomalous as ever.

Out of all this seeming confusion some pattern does emerge: a close relationship of the South American families (with the exception of the erethizontids) and an uncertain mosaic among African forms with most authors excluding pedetids and anomalurids from the group. Ctenodactylids are mainly out and bathyergids hold a shaky

position within the hystricognath–hystricomorph group. But there is an uneasy feeling that African hystricomorphs—whatever they include—have more in common with South American hystricognaths than with any other group of rodents.

The whole group is referred to here as hystricognath-hystricomorph.

THE FOSSILS

Although, in theory, the fossil evidence should provide the answer to the relationship and place of origin of a group of animals there are several reservations about its usefulness. In the case of the hystricognath–hystricomorph rodents, the rocks reveal them fully formed in the lower Oligocene of both South America and Africa.

In South America of 33.6–35.4 million,[1] years ago—the Deseadan stage of the Oligocene—there are already seven families (Hoffstetter and Lavocat 1970; Marshall *et al.* 1977). They were distributed from Bolivia to Patagonia (Fig. 9.2) and already showed considerable variation in, for example, tooth structure. There were woodland browsers with low-crowned (brachyodont) teeth; there were savannah browsers with mid-crowned (mesodont) teeth able to chew up tougher leaves and stalks; and there were grazers on the newly evolving grassland with high-crowned (hypsodont) teeth (Wood an Patterson 1959; Hoffstetter 1968; Hoffstetter and Lavocat 1976; Patterson and Wood 1982; Savage and Russell 1983).

Where did this great variety of forms come from? In the earlier well-marked Argentine Eocene of the Mustersan stage 40–48 million years ago, there are no rodent fossils and yet the rocks abound in marsupials, armadillos, ground sloths, and South American herbivorous ungulates (Simpson 1978, 1980; McKenna 1980). Somewhere in the four million year interval between the Mustersan and the Deseadan stages ancestral rodents must have arrived in South America. The fossil evidence does not reveal when they came or where they came from.

[1] See footnote 1, page 120.

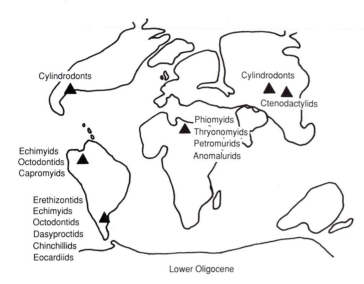

Fig. 9.2 Fossiliferous localities of rodents.

In Africa (Fig. 9.2 and 9.3) two or three definitive hystricomorph families in the Fayum beds of Egypt (approximately 32.5 million years old) had occupied various types of open woodland where they browsed with low-crowned teeth. A few forms like *Gaudeamus* and the later *Neosciuromys* had already developed higher-crowned teeth—like the modern thryonomyid cane rats— and were probably adapting to a coarser vegetation (Hopwood and Hollyfield 1954; Simons 1968; Wood 1968; Cooke 1978; Savage and Russell 1983).

Again, where did these well-differentiated forms come from? There was certainly no abundance of rodents in the African Eocene although there were representatives of the wide-ranging primitive palaeoryctid insectivores (Capetta *et al.* 1978) and flesh-eating creodonts (Savage 1978). There were ancestral elephants (Andrews 1904; Mahboubi *et al.* 1975; Meyer 1978; Sudre 1979), an marsupials (Mahboubi *et al.* 1983) and possibly a primate (Sudre 1979) in North Africa. But unlike the Eocene rocks of South America, those of North Africa have revealed a few rodents. In the Late Eocene of eastern Algeria, teeth of *Protophiomys* and an anomalurid have been found (Jaeger *et al.* 1985). The *Protophiomys* teeth have considerable similarities to those of some of the later Fayum rodents.

Thus there were well-differentiated rodents in

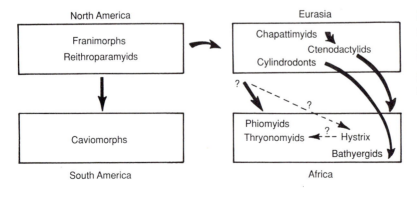

Fig. 9.3 Locations and migration routes of hystricognaths and their relatives during Upper Eocene and Lower Oligocene, according to Patterson and Wood.

the Early Oligocene of both South America and Africa and a few a little earlier in the Late Eocene of North Africa. But where were the ancestors of these two groups? Was there just one ancestor common to the two with the diagnostic hystricognath or hystricomorph characters already developed or were there several less-differentiated ancestors giving rise to the similarities through parallel evolution?

Failing any definitive ancestors in South America or Africa, Wood 1972, 1974) and Wood and Patterson (1959, 1970), and Patterson and Wood (1982) looked for ancestors in the north (Fig. 9.3). But the finds tend to be fragmentary and inferences must be drawn from a few teeth or skulls squashed and shattered in the critical place. Nevertheless, Patterson and Wood decided that the basic characteristics of these rodents is the hystricognathous jaw. As they point out, there is not much to be gained by demanding hystricognathy and hystricomorphy together because there are hystricomorphous skulls that are not hystricognathous (for example SouthAmerican guinea-pigs and African gundis) and protrogomorphous skulls that are hystricognathous (for example Africa mole rats). Given the argument that hystricognathy is the basic feature of the sought-after ancestor (though it may be only a convenient palaeontological character) then all 'incipient hystricognaths' must be considered.

Several examples can be found among generalized early rodents (Wood 1980). Among them, *Reithroparamys* from the Middle Eocene of Texas was protrogomorphous, had three-cusped teeth and showed a tendency to hystricognathy in the position of the angle of the mandible (although Dawson 1977 and Korth 1984 disagree). The only two postcranial skeletons known have long hind legs. The even earlier Eocene *Franimys* from Wyoming might also, according to Wood (1985), be incipiently hystricognathous. From some such ancestors, a family of roller-tooth cylindrodonts appear in the Eocene of North America (some 40–48 million years ago) and a few million years later they can be found in Eurasia (Fig. 3). They may be hystricognathous

(Wood 1984 but Korth 1984 disagrees)—they are certainly protrogomorphous—and they have fourcrested low-crowned teeth.

The early American forms, according to the Patterson and Wood theory, gave rise to fully hystricognathous forms in South America, such as the Patagonian squashed-mouse *Platypittamys*. It had low-crowned four-crested teeth, and as well as being hystricognathous had a slight suggestion of an enlarged infraorbital canal (incipiently hystricomorphous?) *Platypittamys* and its relatives could be ancestral to the South American hystricognaths through animals like the modern, but conservative, octodontid degus. The ancestral group clearly found conditions congenial in South America and, sharpened by competition with the herbivorous ungulates and preyed on by carnivorous marsupials, radiated rapidly into the echimyid spiny-rat group which soon became arboreal (the dactylomyine coro-coros), riverine (capromyine coypus), and fossorial (ctenomyine tuco tucos). At much the same time, a family (Eocardiidae) appeared that later radiated into dasyproctid agoutis and their relatives the chinchillids. By the Miocene at the latest, the caviid guinea-pigs and the hydrochoerid capybaras (Thenius 1976) had emerged from this group.

The New World erethizontid porcupines are present in the Lower Oligocene of Patagonia but, as seen through the fossils, the affinities with other Oligocene hystricognaths are not clear.

Meanwhile, on the other side of the world, the Asian cylindrodonts, according to Patterson and Wood (1982), evolved between 35–29 million years ago into an Oligocene family that was fully hystricognathous, had high-crowned crescent-ridged teeth but was still protrogomorphous. This family, the Tsaganomyidae, were Asian and, according to Patterson and Wood, directly ancestral to the modern bathyergid mole rats which appear for the first time in the African Miocene of 22–17.5 million years ago (Lavocat 1973, 1988). The mole rat fossils come from what was a forested East Africa and a drier Namibia. They were probably already burrowers, digging out loose earth and crushing bulbs and roots with their hystricognathous jaws.

Other possible descendants of the cylindrodonts turn up from 32.5 million years ago in Egypt—in the early Oligocene beds at Fayum—as a varied group of families with fully hystricognathous and hystricomorphous jaws. Among them, *Gaudeamus*, for example, was already, according to Patterson and Wood, like a modern thryonomyid cane rat. The characteristic thryonomyid high-crowned teeth with three crests and encircling ridges was already developing. These animals may have been living in scrub woodland browsing on rough vegetation for which high-crowned teeth and a fully hystricognathous–hystricomorphous grinder jaw was suitable. Alongside them were the equally hystricognathous–hystricomorphous phiomyids with low-crowned fourcrested teeth. Eventually, they gave rise, through later and more southern phiomyids, to the modern petromurid rock rats of arid areas of South-west Africa, today largely grass-eaters. Hystricid Old World porcupine fossils have not been found in the Fayum beds, and Patterson and Wood (1982) and Wood (1985) are of the opinion that there is no way they could have been evolved from the phiomyid–thryonomyid stock. But the fossil history is poorly documented and it is not clear whether they radiated in Asia or Africa. Their affinities are even more elusive than those of the New World porcupines.

So much for the Patterson and Wood Hystricognathi—a monophyletic group of South American and African rodents originating in the North from cylindrodonts and their relatives and then developing in parallel. Patterson and Wood do not include the ctenodactylid gundis, the pedetid spring hares, nor the anomalurid flying squirrels in their Hystricognathi.

A somewhat different view of the same fossils (Fig. 9.4) is taken by Lavocat (1969, 1973, 1974, 1977, 1980) and Hoffstetter (1972, 1980). In their interpretation, the similarities between some fossils of the South American Deseadan Oligocene and some African phiomyids of the same age are so great that they are unlikely to have been developed in parallel from northern ancestors of several million years earlier. Lavocat (1973) points to the near identity of the structure of the tympanic area of the ear in the modern South American vizcacha *Lagostomus* (a chinchillid) and the Miocene African *Diamantomys* (a phiomyid); and to the similarities between the five-crested low-crowned teeth of the lower Oligocene Patagonian *Protosteiromys* (an erethizontid) and the miocene Kenyan *Simonimys* (a phiomyid); and, again, to the similarities between the four-crested teeth of the Egyptian *Gaudeamus* (a thryonomyid) and the Patagonian *Deseadomys* (an echimyid). He lists a further 16 characteristics of the skull in which some members of each group are virtually identical, including of course hystricomorphy and hystricognathy, but also including such characteristics as the position of

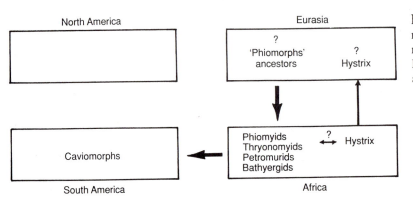

Fig. 9.4 Locations and migration routes of hystricognaths during the Upper Eocene and Lower Oligocene, according to Lavocat.

the opening of the lacrimal or tear ducts, the pattern of articulation at the jaw hinge, and the type of premolar and molar teeth.

According to Lavocat, the modern African thryonomyid cane rats and the petromurid rock rats are direct descendants of these early phiomyids. The bathyergid mole rats, too, are on a direct line from the early phiomyids. This view, following Stromer (1923), necessitates a modification of an already acquired hystricomorphy to the weak hystricomorphy — near protrogomorphy — of modern bathyergids. Lavocat finds this an acceptable adaptation to a fossorial life particularly as some phiomyids had only small infraorbital canals. This seems the more likely since the finding of a new Miocene bathyergid fossil in Kenya which, according to Lavocat, showed signs of regressive hystricomorphy (Lavocat 1988). Furthermore, it has been found that embryonic bathyergids have small muscle bundles of the anterior part of the masseter medialis which then degenerate in most species (Maier and Schrenk 1987)—again suggesting regression from a hystricomorphous state. The hystricid porcupines remain a puzzle. A tooth found in the lower Miocene beds of Pakistan is attributed by Black (1972) to *Hystrix*, but the first confirmed fossil Old World porcupines are from Mid-Miocene Pakistan and Turkey (Raza *et al.* 1983, Savage and Russell 1983). Lavocat believes the hystricids are closely related to the thryonomyids and derived from an early phiomyid (the opposite evolutionary direction from that of Patterson and Wood).

Lavocat includes the pedetid spring hares and anomalurid flying squirrels in this African group, but already anomalurids seem to have been gliding through the trees in the Late Eocene and the Oligocene (Jaeger *et al.* 1985) and, by the Miocene, the pedetids were browsing and grazing in East Africa Lavocat is no more certain of their immediate ancestry than anyone else. And the ctenodactylids seem to be late migrants to Africa and not, according to Lavocat, easy to ally with the phiomyid derivatives.

These two opinions—one advocating the origin of all hystricognaths from northern cylindrodonts

and their relatives and an early dichotomy to give separate New World and Old World groups, and the other suggesting an Asian or African origin from unknown ancestors an a much later dichotomy in Asia or Africa—may be modified as more fossils are found.

For example, several teeth have been found in the Upper Eocene of eastern Algeria belonging to a primitive phiomyid which Coiffait (1984) and Jaeger (1985) and their colleagues claim to belong to a type that could be ancestral to the later Egyptian phiomyids. The *Protophyiomys* teeth are generally like Oligocene phiomyid teeth but unfortunately the jaws have not been found so it is impossible to know whether they were hystricognath and hystricomorph like the phiomyids.

In 1978, Hussain *et al.* described rodent fossils from MidEocene Pakistan that were hystricomorphous but not hystricognathous. They called them the Chapattimyidae and suggested they were ancestors of the African phiomyids on the one hand and North American cylindrodont rodents on the other.[1] These two groups would subsequently give rise to the separate New and Old World hystricognaths in parallel. And it was realized that these chapattimyids resemble Eocene ctenodactylids (Dawson 1964, Shevyreva 1972, so closely that they can be united into a ctenodactyloid group.

Now Jaeger *et al.* consider their new *Protophyiomys* shares many characters with the chapattimyids and may have been derived from some ancestral chapattimyid that migrated to Africa as Hussain *et al.* had tentatively suggested. Furthermore, they regard the hystricid porcupines as the living representatives of the Asian chapattimyids. The evidence for this is slim.

Thus, there is some support for an Asian hystricomorphous ctenodactyloid as the ancestor of all hystricognaths.

Meanwhile Dawson *et al.* (1984) described a rodent called *Cocomys* which, although protrogomorphous, had a slightly enlarged infraorbital canal indicative of a developing hystricomorphy.

[1] These authors later abandoned this idea that the Chapattimydae might be ancestors of the American forms (R.L.).

Cocomys is the oldest known Asian rodent from the Lower Eocene of China and can be closely associated with the ctenodactyloids.

The picture that is emerging is of an early radiation of ctenodactyloids in Asia from which the African and South American hystricognath–hystricomorph rodents evolved. Ctenodactylids and possibly hystricids may be the living representatives of this early radiation. It is reasonable to assume that these early ctenodactyloids migrated to Africa but it is more difficult to guess how they got to South America. The fossil evidence gives no clues.

A NUMERICAL APPROACH

The analysis of present-day can take into account a wide range of characters although most authors have restricted themselves to comparatively few. A large number of characters permits the use of statistical analyses. From various sources I collected 69 characteristics of the ten South American and seven African families shown in Table 9.1 together with most other 'traditional' rodent families, and subjected them to a cluster analysis. The method chosen was a graph linkage variant of single line cluster analysis (Estabrook 1966, Bisby 1973).

The characters include skull and postcranial skeleton (Tullberg 1899; Ellerman 1940; Wood 1974); muscles (Woods 1972, 1975; Wood 1974); structure of the ear (Landry 1957; Parent 1976, 1980); structure of various internal organs, for example stomach, brain, and lung lobes (Toepfer 1891; Tullberg 1899; Franklin 1948; Pilleri 1960; Carleton 1973; Gunderson 1976); patterns of the blood vascular system (Bugge 1971, 1974; George 1981); structure of the male and female reproductive organs (Tullberg 1899; Didier 1956, 1962, 1965; Burt 1960; Asdell 1964; Weir 1974; George 1978*a*; Butynski 1979; Eisenberg 1981); embryology (Luckett 1971, 1980); hair structure (Hausman 1920; Mayer 1952; George 1978*b* and personal observation); foot pad pattern (Tullberg 1899; Pocock 1922;

Patton and Gardner 1972; Grand and Eisenberg 1982); karyotypes (Hsu and Benirschke 1977; Matthey 1956, 1958; Nadler *et al.* 1975; George and Weir 1974; George 1980); and zoogeographical distribution. Parasite distribution (Durette-Desset 1971; Quentin 1973) and biochemical evidence (Beintema 1985, for example) were not included as they did not cover enough families. In addition, a considerable amount of information on soft parts is taken from personal observation.

The cluster analysis provided linkage groups (Fig. 9.5). It demonstrates a very close linkage of South American octodontids, echimyids, dasyproctids, capromyids, and hydrocherids—with ctenomyids, caviids, chinchillids, and African thryonomyids only a little less closely linked. Petromurids link with thryonomyids. And at about the same level hystricids and erethizontids link independently with the South American families. At a lower level, bathyergids link with the main group and, finally, ctenodactylids link to thryonomyids and petromurids. Anomalurids and pedetids are as close to the myomorph-sciuromorph group as they are to the hystricognath–hystricomorph group, although they have some similarities to one another. The main conclusions of such an analysis—with no weighting of characters—is that rodents fall into a group of myomorph–sciuromorphs and a group of hystricognath–hystricomorphs with the anomalurids and pedetids hovering in the background. Within the hystricognath–hystricomorphs must be included, though distantly, both ctenodactylids and bathyergids (they bear no overall resemblance to myomorphs or sciuromorphs). Thryonomyids are very like the South American families but erethizontids and hystricids rather less so. This is a grouping of like with like, but it does not provide a phylogeny because it gives no weight to characters and it does not distinguish characters derived through a common ancestor, and those acquired in parallel. It does not establish a monophyletic line but it does give a useful general grouping of families and it provides a basis for asking questions.

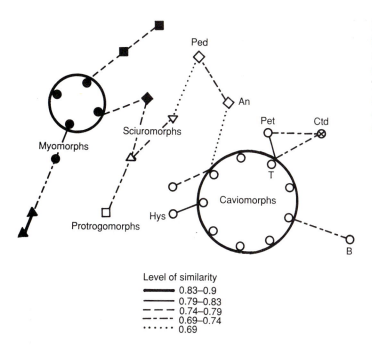

Fig. 9.5 Linkage diagram for rodent families using 69 character states. Black symbols, myomorphs; ▽, sciuromorphs; □, protrogomorphs; ○, hystricognaths; T, thryonomyids; Ped, pedetids; An, anomalurids; Pet, petromurids; Ctd, ctenodactylids; B, bathyergids; Hys, hystricids. (After George, 1985.)

Level of similarity
— 0.83–0.9
— 0.79–0.83
– – – 0.74–0.79
–·–·– 0.69–0.74
········ 0.69

A further analysis of the families used cladistic methods; 31 shared characters were charted along the lines suggested by Hennig (1966), Nelson (1979), and Wiley (1981). In this type of analysis, the problem is to decide which characters are derived or specialized and which are not, without assuming a phylogeny first. The criterion of choice was the generality of the character within the rodents, the less general being regarded as specialized, derived, or apomorphic. However, a good deal of subjective judgment is involved and characters that seem 'useful' tend to be selected. The result of this analysis was, again, an indication of a hystricognath–hystricomorph group and a myomorph-sciuromorph group, with the anomalurids and pedetids in between. Then characters shared with any myomorph–sciuromorph were eliminated because it was difficult to decide whether they were independently derived or a hangover from shared ancestors, and this eliminated hystricomorphy which occurs in the dipodid–zapodid families. Many authors would derive myomorphy from hystricomorphy, perhaps on more than one occasion (Lindsay 1977; Vianey-Liaud 1985). Table 9.3 lists the remaining 20 characters (out of 31) and Figure 9.6 the groupings derived from them., The order in the figure is determined by the concept that there should be as few changes of character states as possible.

The supposedly derived shared characters are not obviously closely correlated either functionally or embryologically although they do concern mainly the hard structures, muscles, and reproductive systems. The families and the characters can be arranged in different ways but whichever method of assessment is used, the same myomorph–sciuromorph and hystricognath–hystricomorph blocks emerge. The cladogram is satisfactorily similar to the one produced by Luckett (1985) using a different set of 20 character states—mostly embryological.

Within the hystricognath–hystricomorph block itself (now with only 12 character states utilized) there is a bunch of South American families with hystricids and erethizontids close

Table 9.3 Selected characters for a phylogenetic analysis of rodent families (after George, 1985)

General	Derived
1. Malleus and incus free	Malleus and incus fused
2. Uniserial enamel on teeth	Multiserial enamel
3. Vagina patent	Vaginal closure membrane
4. Penis simple	Penis with saccula urethralis
5. Parietal endoderm of blastocyst complete	Parietal endoderm incomplete
6. No scapuloclavicularis muscle	Scapuloclavicularis muscle
7. Nipples on ventral mammary line	Lateral nipples
8. Chromosomes uniform	Chromosomes with big n.o.r.
9. Subplacenta absent	Subplacenta present
10. Masseter lateralis profundus undivided	Masseter lateralis profundus divided
11. Implantation of embryo superficial	Implantation interstitial
12. Sciurognathy	Hystricognathy
13. Stylohyoideus muscle absent	Stylohyoideus muscle present
14. Lung lobes 2–4:3	Lung lobes 1:3
15. Tibia and fibula free	Tibia and fibula fused
16. No entepicondylar foramen in humerus	Entepicondylar foramen
17. Lacrimal duct inside orbit	Lacrimal duct outside orbit
18. Stapedial artery present	Stapedial artery absent
19. Sciuromorphy	Myomorphy
20. Two or right azygous vein	Left azygous vein

and thryonomyids, petromurids, and bathyergids alongside (Fig. 9.7). Ctenodactylids belong, although more distantly.

Both the cluster analysis and the cladistic analyses suggest close similarity between South American and African hystricognaths.[1] Both analyses include, unequivocally, the bathyergids and ctenodactylids within hystricognath–hystricomorph block—the ctenodactylids as a sister group to the rest. The pedetids occupy a position closer to the hystricognath–hystricomorph group in the cladistic analysis than in the cluster analysis, but the anomalurids show the reverse.

---

[1] Martin (1990) has provided new evidence from his study of the enamel structure of the incisors (R.L.).

PUTTING TWO AND TWO TOGETHER

The interpretation of the fossils led to three possible conclusions. Either the hystricognath-hystricomorph were evolved in parallel from Early Eocene northern rodents—such as *Reithroparamys* and the cylindrodonts—with protrogomorphous jaws which may have been incipiently hystricognath; or they arose from Asian chapattimyid stock with hystricomorphous-sciurognathous jaws. These ancestors spread across the north and then migrated south to South America and Africa. Or they evolved from a rodent like *Cocomys* (protrogomorphous but with a fairly big infraorbital canal) in Eocene China which gave rise to a vast radiation of chapattimyids and ctenodactylids (for example the hystricomorphous *Tamquammys*) in the Eocene (Dawson *et al.* 1984). They and their descendants gradually

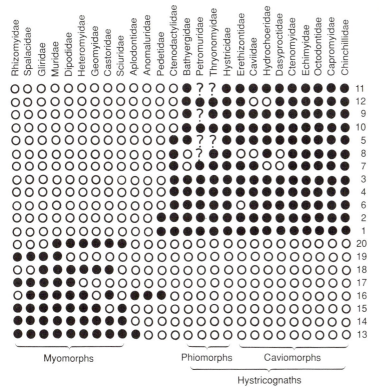

Fig. 9.6 Character phylogeny of rodent families for the 20 characters listed in Table 9.3. Derived character states (see Table 9.3) are indicated by filled circles. (After George 1985.)

spread over Asia and found their way during the Eocene to Africa and thence to South America.

The analysis of the affinities of modern rodent families seems to lend support to this last view. Modern ctenodactylids appear to be a sister group to all hystricognathous rodents (George 1985; Luckett 1985). Thus, with the added evidence of the fossils, ctenodactyloids could represent one of the earliest rodent radiations. It could be supposed that as global temperature dropped and with it global precipitation giving rise to more xeric vegetation (Singh 1988) the rodents developed a hystricognathous jaw better able to chop and grind (Vianey-Liaud 1985). They became adapted to a pastoral life in the open, where seasons were not marked and rains were of short duration. They did not burrow but gave birth, as they always had done, to a few well-developed young at regular intervals. In the open, the mother

remained alert while she suckled her young from lateral nipples. Hearing was adapted to the dry air. It can be supposed that all these were characteristics of the early radiation as they are present in modern ctenodactylids and hystricognaths. Later, improvements occurred in the arrangements for the development of the embryo (Luckett 1985). It is just possible, according to Jaeger *et al.*, that the Old World porcupines had already started to develop from a chapattimyid stock, but the similarities of the hystricids to modern hystricognath-hystricomorphs makes it more likely that they were a somewhat later offshoot.

The protophiomyids differentiated and eventually spread as thryonomyids, petromurids, and South American hystricognath-hystricomorphs. Eventually the bathyergids took to burrowing.

The problem remains of where all this was happening and whether these typical characters of

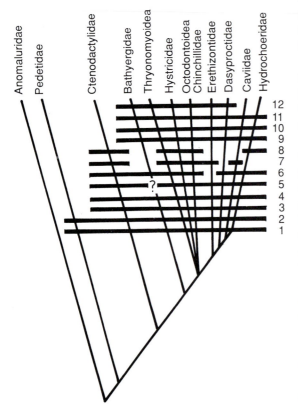

Fig. 9.7 Cladogram of the hystricognath rodents to show their relationships with the Ctenodactylidae, Pedetidae, and Anomaluridae. Numbers refer to the characters listed in Table 9.3. (After George, 1985.)

hystricognath–hystricomorph (such as reproductive arrangements, chewing adaptations, hearing adaptations, and even methods of locomotion) were evolved early and once only, or whether they were evolved at least twice in parallel.

VICARIANCE OR DISPERSAL

It is difficult to square any of the evolutionary theories with the geological facts. The biogeographer considers distribution to be brought about either by vicariance, in which a once widespread population is broken up by geological events, or by dispersal, in which organisms have spread by surmounting such barriers as seas or mountains. There is, of course, no reason why the two should not be combined.

Are the South American and African families the isolated remains of a once world-wide population that was separated by the drift of continents? Or did the families migrate across continents and seas?

Neither the fossils nor the geological facts fit the first (Hershkovitz 1977) extreme vicariant model. Gondwana had fragmented by the mid-Cretaceous (Cenomanian) of 100 million years ago and both South America and Africa were becoming island continents. They had lost contact with one another and with the north by 80 million years ago at the latest (see Chapter 2). The rodents—*Reithroparamys* of North America and *Cocomys* of Asia among them—are unlikely to be much more than 56–58 million years old. It is not until about 39 million years ago that the earliest protophiomyid occurred in Africa and 36 million[1] years ago that hystricognath–hystricomorphs are found in South America. Apart from rodents, monkeys, and a few marsupials (Mahboudi *et al.* 1983, Brown and Simons 1984), the mammalian fossils of the two southern continents are very different from one another in the Oligocene. In South America there was a great variety of carnivorous marsupials, armadillos, and herbivorous notoungulates and ground sloths, while in Africa the elephant stock and hyraxes predominated. As a generalization, vicariant patterns (Croizat 1958; Croizat *et al.* 1974; Nelson and Platnick 1981) do not seem to explain rodent distribution in the Oligocene. The drift of the continents was too early.

If geomorphological changes do not by themselves account for past and present hystricognath–hystricomorph distribution, then the animals must have moved about. They must have dispersed. But where did they start their wanderings and which way did they wander?

See footnote 1, page 120.

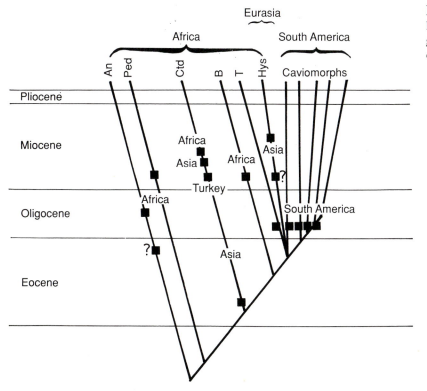

Fig. 9.8 Geographical localities of the rodent groups included in the cladogram of Fig. 9.7.

According to the fossil evidence, Patterson and Wood advocate a fanned migration from the north. Northern cylindrodonts and their relatives migrated from North America—but no hystricomorphous or genuinely hystricognathous fossils have been found in North America[1]—to South America and from Asia to Africa. Meeting similar vegetation and climate in the two southern continents, they evolved most of their hystricognath–hystricomorph characteristics in parallel. A similar story is told by Hussain *et al.* except that the stock starts from Asian chapattimyids and cylindrondonts. According to Lavocat, a migration from Asia to Africa took place as the ancestors gradually evolved into protophiomyids. From there, a few migrants managed to get to South America. The finding of Late Eocene

protophyomids in eastern Algeria adds weight to Lavocat's view.

Analysis of present-day family distribution is not conclusive either. Using a cladistic biogeographical approach, a cladogram of geographical distribution can be constructed (Nelson and Platnick 1981). Figure 9.8 assumes monophyly of the hystricognath-hystricomorph block and assumes the ctenodactylids and bathyergids belong. It is a simplified version of the phylogenetic cladogram of Fig. 9.7. It does not show which way the animals moved but it does alter the possibilities. It establishes ctenodactylid relationships with ancestral hystricognath–hystricomorphs. Ctenodactylids (Wood 1977) and their relatives, the chapattimyids, were in Asia in the Mid Eocene long before any other hystricomorphous or hystricognathous rodents were on the scene. Thus, it seems likely that the hystricognath–hystricomorph had an

[1] *Protoptychus* is reported to be hystricomorphous (R.L.).

Asian origin in the Eocene. They could have diverged at this stage: one group going to the Americas and the other to Africa (a similar interpretation to that of Patterson and Wood 1982 and Hussain *et al.* 1978, but at a later evolutionary stage). But because of the close affinities of the thryonomyid–petromurid families and the South American families it seems more likely that the divergence of the ancestral line did not take place until later. The protophiomyid grade might have been reached in Asia or in Africa in the Eocene, but it was not until it reached Africa that the divergence into the main African families and the ancestor of the South American families took place. This analysis supports Lavocat.

The cladogram of Fig. 9.8 can be superimposed on a map of the world in the Upper Eocene (fol-

lowing Ross 1974). The most parsimonious way of doing this is shown in Fig. 9.9. Again there is an indication that because of the close similarities of some of the African families and the South American families the most likely route was Asia to Africa to South America.

All the dispersal theories require crossing continents and seas. In the Patterson and Wood hypothesis, ancestral franimorphs traversed the northern continents in Early Eocene days. There were interchange corridors of land across Beringia and Greenland in the Lower Eocene. Many mammals used the routes. The American migrants then had a long seacrossing in South America across the western extension of the Tethys Sea and the currents were not favourable, flowing east to west (see Chapter 3). The sea

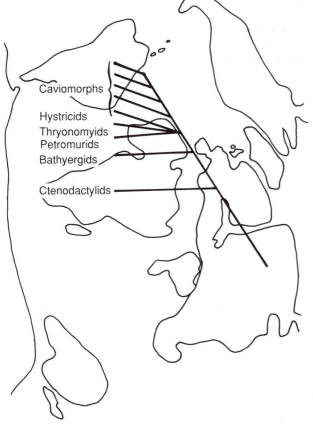

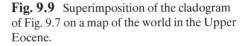

Fig. 9.9 Superimposition of the cladogram of Fig. 9.7 on a map of the world in the Upper Eocene.

Caviomorphs

Hystricids
Thryonomyids
Petromurids
Bathyergids

Ctenodactylids

seems to have existed since the Late Palaeocene or very Early Eocene (White 1986). No other mammal—except possibly the monkeys—made it, although the distance may have been as little as 500 km (Tarling 1980) with a scattering of islands. Webb (1978) argues that the sea-crossing was possible on the grounds that the land tortoise *Geochelone* crossed the barrier. Rodents have a good reputation for waif dispersal (Simpson 1978)—they are the only terrestrial placental mammals, except for the bats, to have reached Australia unaided by man—but even rodents have not colonized mid-oceanic islands to the extent that land tortoises have. On the other side of the world, a Eurasian migration to Africa required a less formidable sea-crossing. It would have been short during an eastern Tethyan regression with the currents helpfully flowing east to west. Probably there were islands, too. Certainly the marsupials (Mahboubi *et al.* 1983) and amphibious pig-like anthracosaurs and primates made the crossing, for example (Coryndon and Savage 1973).

In the Lavocat hypothesis, the Atlantic sea-crossing is the alternative to the North America–South America crossing (both hypotheses accept the Asia–Africa sea-crossing). The prevailing ocean circulation could help a voyage from West Africa to the Brazilian coast. The ocean was still warm (Kennet 1981). There may have been 'stepping stone' islands on the Sierra Leone and Ceará Rises, resulting in only 200–300 km water gaps in the 1000–1500 km of the Atlantic (Funnel and Smith 1968, Tarling 1980) between the African and South American coasts. The Atlantic crossing has been favoured by several authors (Lavocat 1969, 1977, 1980; Hoffstetter 1972, 1975, Ciochon and Chiarelli 1980).

I favour the Atlantic crossing because of the seemingly Asian origin of the hystricognath–hystricomorph block and the late divergence of the African and South American families. The Eurasia–Africa–South American migration traverses only 80° of latitude, keeping the animals within a present-day 30°N and 50°S. The Asia–North America–South American migration takes them through 140 latitude, through a narrowing Beringia—with a climate becoming markedly seasonal and variable (Napier 1971, Keller and Barron 1983)—and down the length of North America. The greater the latitudinal migration the greater the climatic and vegetational variation and the greater the variation that might be expected in the animals. Yet the hystricognath–hystricomorph of the two southern continents are very similar and predominantly adapted to tropical and subtropical open habitats, to browsing and grazing on coarse vegetation. (Fig. 9.10)

Whatever happened, dispersal seems inevitable. Whichever way they went, they crossed 20 000 km. But they had at least five million years to do it, which is only four metres a year.

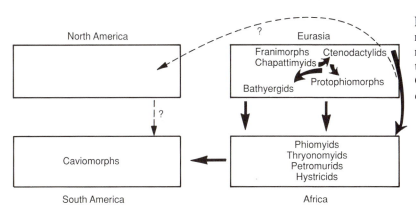

Fig. 9.10 Locations and migration routes of hystricognaths and their relatives during the Upper Eocene and Lower Oligocene, as suggested in this chapter.

REFERENCES

Andrews, C. W. (1904). Further notes on the mammals of the Eocene of Egypt, *Geological Magazine*, decade 5, **1**, 157–62.

Asdell, S. A. (1964). *Patterns of mammalian reproduction.* Cornell University Press.

Beintema, J. J. (1985). Amino acid sequence data and evolutionary relationships among hystricognath and other rodents. In *Evolutionary relationships among rodents. A multidisciplinary analysis.* (ed. W. P. Luckett and J. L. Hartenberger), pp. 549–65. Plenum Press, New York.

Beecher, R. M. (1979). Functional significance of the mandibular symphysis. *Journal of Morphology*, **159**, 117–30.

Bisby, F. A. (1973). The role of taximetrics in Angiosperm taxonomy. *New Phytology*, **72**, 699–726.

Black, C. C. (1972). Review of fossil rodents from the Neogene Siwalik beds of India and Pakistan. Paleontology, **15**, 238–66.

Brandt, J. F. (1855). Beiträge zur nähern Kenntniss der Säugethiere Russlands, *Mémoires de l'Académie Impériale des Sciences, St Petersbourg*, sér. 6, **9**, 1-365.

Bown, T. M. and Simons, E. L. (1984). First record of Marsupials (Metatheria, Polyprotodonta) from the Oligocene in Africa. *Nature*, London. **308**, 447–9.

Bugge, J. (1971). The cephalic arterial system in New and Old World hystricomorphs and in bathyergids, with special reference to the classification of rodents. *Acta Anatomica*, **80**, 516–36.

Bugge, J. (1974). The cephalic arterial system in insectivores, primates, rodents and Iagomorphs with special reference to the systematic classification. *Acta Anatomica*, **87**, (suppl. 62), 1–160.

Burt, W. H. (1960). Bacula of North American mammals. *Miscellaneous Publications of the Museum of Geology*, University of Michigan, **113**, 1–175.

Butynski, T. M. (1979). Reproductive ecology of the Springhaas *Pedetes capensis* in Botswana. *Journal of Zoology*, **189**, 221–32.

Byrd, K. E. (1981). Mandibular movement and muscle activity during mastication in the guinea pig (*Cavia porcellus*). *Journal of morphology*, **170**, 147–69.

Cappetta, E., Jaeger, J. J., Sabatier, M., Sigé, B., Sudre, J., and Vianey-Liaud, M. (1978). Découverte dans le Paléocène du Maroc des plus anciens mammifères euthériens d'Afrique. *Geobios*, **11**, 257–62.

Carleton, M. (1973). A survey of gross stomach morphology in New World Cricetinae (Rodentia: Muridea) with comments on functional intepretations. *Miscellaneous Publications of the Museum of Zoology*, University of Michigan, **146**, 1–46.

Chaline, J. and Mein, P. (1979). *Les rongeurs et l'évolution.* Doin, Paris.

Ciochon, R. L. and Chiarelli, A. B. (ed.) (1980). *Evolutionary biology of the New World monkeys and continental drift*, pp. 1–528. Plenum Press, New York.

Coiffait, P. E., Coiffait, B., Jaeger, J. J., and Mahboubi, M. (1984). Un nouveau gisement à mammifères fossiles d'âge Eocène supérieur sur le versant sud des Nementcha (Algérie orientale): découverte des plus anciens rongeurs d'Afrique. *Comtes Rendus de l'Académie des Sciences*, Paris, **299**, 893–8.

Cooke, H. B. S. (1978). Africa: the physical setting. In *Evolution of African mammals* (ed. V. J. Maglio and H. B. S. Cooke), pp. 17–45. Harvard University Press, Cambridge, Massachusetts.

Coryndon, S. C. and Savage, R. J. C. (1973). The origin and affinities of African mammal faunas. Organisms and continents through time. *Special paper in paleontology*, no. 12, pp. 121–35. Systematic Association Publications.

Croizat, L. (1958). *Panbiogeography.* 3 vol., Wheldon and Wesley, Codicot, England.

Croizat, L., Nelson, G., and Rosen, D. E. (1974). Centers of origin and related concepts. *Systematic Zoology*, **23**, 265–87.

Dawson, M. R. (1964). Late Eocene rodents (Mammalia) from inner Mongolia. American Museum Novitates, **2191**, 1–15.

Dawson, M. R. (1977)., Late Eocene rodent radiations in North America, Europe and Asia. *Geobios*, Mémoire spécial, **1**, 195-209.

Dawson, M. R., Li, C. K., and Qi, T. (1984). Eocene ctenodactyloid rodents (Mammalia) of eastern and central Asia. In *Papers in vetebrate paleontology honoring R. W. Wilson* (ed. R. Mengel), Special publications of the Carnegie Museum of Natural History, **9**, pp. 138–50.

Didier, R. (1956). Etude systématique de l'os pénien des mammifères. *Mammalia*, **20**, 238–47.

Didier, R. (1962). Note sur l'os pénien de quelques rongeurs de l'Amérique du Sud. *Mammalis*, **26**, 408–30.

Didier, R. (1965). Etude systématique de l'os pénien des mammifères. *Mammalia*, **29**, 331–42.

Durette-Desset, M. C. (1971). Essai de classification

des Nématodes Héligmosomes. Corrélations avec la paléobiogéographie des hôtes. *Mémoires du Muséum National d'Histoire Naturelle, Paris*, **69**, 1–136.

Eisenberg, J. F. (1981). *The mammalian radiations.* Athlone Press, London.

Ellerman, J. R. (1940). *The families and genera of living rodents.* Vol. 1, 2, British Museum (Natural History).

Estabrook, G. F. (1966). A mathematical model in graph theory for biology classification. *Journal of Theoretical Biology*, **12**, 297–310.

Franklin, K. J. (1948). *Cardiovascular studies.* Oxford.

Funnel, B. M. and Smith, A. G. (1968). Opening of the Atlantic Ocean. *Nature*, London. **219**, 1328–33.

George, W. (1978*a*). Reproduction in female gundis (Rodentia) (Ctenodactylidae). *Jounal of Zoology*, **185**, 57–71.

George, W. (1978*b*). Combs, fur and coat care in the etenodactylidae (Rodentia). *Zeitschrift für Säugetierkunde*, **43**, 143–55.

George, W. (1980). A study in hystricomorph rodent relationships: the karyotypes of *Thryonomys gregorianus. Pedetes capensis* and *Hystrix cristata. Zoological Journal of the Linnean Society*, **68**, 361–72.

George, W. (1981). Blood vascular patterns in rodents. Contribution to an analysis of rodent family relationships. *Zoological Journal of the Linnean Society*, **73**, 287–306.

George, W. (1985). Reproductive and chromosomal characters of Ctenodactylids as a key to their evolutionary relationships. In *Evolutionary relationships among rodents. A multidisciplinary analysis* (ed. W. P. Luckett and J. L. Hartenberger), pp. 453–74. Plenum Press, New York.

George, W. and Weir, B. (1974). Hystricomorph chromosomes. In *The biology of Hystricomorph rodents*, Symposia of the Zoological society of London, No. 34. (ed. I. W. Rowlands and B. J. Weir), pp. 143-60. Academic Press, London.

Gevin, P., Lavocat, R., Mongereau, N., and Sudre, J. (1975). Découverte de mammifères dans la moitié inférieure de l'Eocène continental du Nord-Ouest du Sahara. *Comptes Rendus de l'Académie des Sciences, Paris*, **280**, 967–8.

Grand, T. I. and Eisenberg, J. F. (1982). On the affinities of the Dinomyidae. *Säugetierkunde Mitteilungen*, **30**, 151–7.

Gunderson, H. L. (1976). *Mammalogy*, McGraw-Hill, New York.

Hausman, L. A. (1920). Structural characteristics of the hair of mammals. *American Naturalist*, **54**, 496–523.

Hennig, W. (1966). *Phylogenetic Systematics*. University of Illinois Press, Urbana.

Herschkovitz, P. (1977). *Living New World monkeys (Platyrrhini)*, Vol. 1, Chicago University Press.

Hiiemae, K. M. (1978). Mammalian mastication. A review of the activity of the jaw muscles and the movements they produce in chewing. In *Development, function and evolution of the teeth* (ed. P. M. Butler and K. A. Joysey), pp. 359–98. Academic Press, London.

Hoffstetter, R. (1968). Un gisement de mammifères déséadiens (Oligocène inférieur) en Bolivie. *Comptes Rendus de l'Académie des Sciences, Paris.* **267D**, 1045–97.

Hoffstetter, R. (1972*a*). Origine et dispesion des rongeurs hystricomorphes. *Comptes Rendus de l'Académie des Sciences, Paris.* **274D**, 2867–70.

Hoffstetter, R. (1972*b*). Relationships, origins and history of the ceboid monkeys and caviomorph rodents: a modern reinterpretation. *Evolutionary Biology*, **6**, 323–47.

Hoffstetter, R. (1975). El origen de los Caviomorpha y el problema de los Hystricognathi (Rodentia). I Congreso Argentino de paleontologia y bioestratigrafia, *Actas* no. 1, **Tome 2**, pp. 505–28.

Hoffstetter, R. (1980). Origin and development of New World monkeys emphasizing the southern continents route. In *Evolutionary biology of the New World monkeys and continental drift* (ed. R. L. Ciochon and A. E. Chiarelli), pp. 103–22, Plenum Press, New York.

Hoffstetter, R., and Lavocat, R. (1970). Découverte dans le Déséadien de Bolivie de genres pentalophodontes appuyant les affinités africaines des rongeurs caviomorphes. *Comptes Rendus de l'Académie des Sciences, Paris.* **271**, 172–5.

Hoffstetter, R. and Lavocat, R. (1976). Rongeurs caviomorphes de l'Oligocène de Bolivie. *Palaeovertebrata*, **7**, (3), 1–90.

Hopwood, A. T. and Hollyfield, J. P. (1954). An annotated bibliography of the fossil mammals of Africa (1472–1950). In *Fossil mammals of Africa*, vol. **8**, pp. 1–194. British Museum (Natural History).

Hsu, T. C. and Benirschke, K. (1977). *An atlas of mammalian chronosomes*, Vol. 10, Springer-Verlag, Berlin.

Hussain, S. T., Bruijn, H. de., and Leinders, J. M.

(1978). Middle Eocene rodents from the Kala Chitta range (Punjab, Pakistan). *Proceedings of the Koninklijke Nederlandse Akademie van Wetenschappen*, series B, **81**, 74–112.

Jaeger, J. J., Denys, C., and Coiffait, B. (1985). In *Evolutionary relationships among rodents. A multidisciplinary analysis* (ed. W. P. Luckett and J. L. Hartenberger), pp. 567–88, Plenum Press, New York.

Keller, G. and Barron, J. A. (1983). Paleoceanographic implications of Miocene deep-sea hiatuses. *Bulletin of the Geological Society of America*, **94**, 590–613.

Kennett, J. (1981). *Marine Geology*. Prentice-Hall, New Jersey.

Kesner, M. H. (1980). Functional morphology of the masticatory musculature of the rodent subfamily Microtinae. *Journal of Morphology*, **165**, 205–22.

Korth, W. W. (1984). Earliest tertiary evolution and radiation of rodents in North America. *Bulletin of the Carnegie Museum Natural History*, **24**, 1–71.

Landry, S. O. (1957). The interrelationships of the New and Old World hystricomorph rodents. *University of California Publications in Zoology*, **56**, 1–118.

Lavocat, R. (1969). La systématique des rongeurs hystricomorphes et la dérive des continents. *Comptes Rendus de l'Académie des Sciences, Paris*, **269**, 1496–7.

Lavocat, R. (1973). Les rongeurs du Miocène d'Afrique orientale I. Miocène inférieur. *Mémoires et Travaux de l'E.P.H.E., Institut de Montpellier*, **1**, 1–284.

Lavocat, R. (1974a). The interrelationships between the African and South American rodents and their bearing on the problem of the origin of South American monkeys. *Journal of human evolution*, **3**, 523–6.

Lavocat, R. (1974b). What is an Hystricognath? In *The biology of Hystricomorph rodents, Symposia of the Zoological society of London, No. 34*, (ed. I. W. Rowlands and B. J. Weir), pp. 7–20, 55–60, Academic Press, London.

Lavocat, R. (1977). Sur l'origine des faunes sud-américaines de mammifères du Mésozoïque terminal et du Cénozoïque ancien. *Comptes Rendus de l'Académie des Sciences, Paris*. **285**, 1423–6.

Lavocat, R. (1980). The implications of Rodent paleontology and biogeography to the geographical sources and origin of Platyrrhine Primates. In *Evolutionary biology of the New World monkeys and continental drift* (ed. R. L. Ciochon and A. B. Chiarelli), pp. 93–102, Plenum Press, New York.

Lavocat, R. (1988). Un rongeur bathyergidé nouveau remarquable du Miocène de Port Ternan. *Comptes Rendus de l'Académie des Sciences, Paris*. **306**, 1301–4.

Lindsay, E. H. (1977). *Simimys* and the origin of the Cricetidae (Rodentia, Muridae) *Geobios*, **10**, 597–623.

Luckett, W. P. (1971). The development of the chorio-allantoid placenta of the african scaly-tailed squirrels (family Anomaluridae). *American Journal of Anatomy*, **130**, 159–78.

Luckett, W. P. (1980a). Fetal membranes and placental development in the African hystricomorphous rodent *Ctenodactylus. Anatomical Record*, **196**, 116A.

Luckett, W. P. (1980b). Monophyletic or diphyletic origin of Anthropoidea and Hystricognathi: evidence of the fetal membranes. In *Evolutionary biology of the New World monkeys and continental drift* (ed. R. L. Ciochon and A. B. Chiarelli), pp. 347–68. Plenum Press, New York.

Luckett, W. P. (ed.) (1985). Superordinal and intra-ordinal affinities of rodents: developmental evidence from the dentition and placentation. In *Evolutionary relationships among rodents. A multidisciplinary analysis* (ed. W. P. Luckett and J. L. Hartenberger), pp. 227-76. Plenum Press, New York.

Macfadden, B. J., Campbell, K. E., Cifelli, R. L., Siles, O., Johnson, N. M., Naeseer, C. W., and Zeitler, P. K. (1985). Magnetic polarity stratigraphy and mammalian fauna of the Deseadan (Late Oligocene–Early Miocene) Salla beds of Northern Bolivia. *The Journal of Geology*, **93**, 223–250.

McKenna M. C. (1980). Early history and biogeography of South America's extinct land mammals. In *Evolutionary biology of the New World monkeys and continental drift* (ed. R. L. Ciochon and A. B. Chiarelli), pp. 43–77, Plenum Press, New York.

Mahboubi, M., Ameur, R., Crochet, J. Y., and Jaeger, J. J. (1983). Première découverte d'un marsupial en Afrique. *Comptes Rendus de l'Académie des Sciences*, **297**, 691–4.

Mahboubi, M., Ameur, R., Crochet, J. Y., and Jaeger, J. J. (1984). Earliest known proboscidian from early Eocene of North West Africa. *Nature*, London. **308**, 543–4.

Maier, W., and Schrenk, P. (1987). The hystricomorphy of the Bathyergidae, as determined from ontogenetic evidence. *Zeitschrift fur Säugetierkunde*, **52**, 156–64.

Marshall, L. G., Pascual, R., Curtis, G. H., and Drake, R. E. (1977). South American geochronology: Radiometric time scale for middle to late Tertiary mammal-bearing horizons in Patagonia. *Science*, **195**, 1325–8.

Martin, T. ()1990). Origin of the caviomorphs: evidence from incisor enamel, *Journal of Vertebrate Paleontology*, **10**, (3) (Abstracts), 34A.

Matthey, R. (1956). Nouveaux apports à la cytologie comparée des rongeurs. *Chromosome*, **7**, 670–92, Berlin.

Matthey, R. (1958). Les chromosomes des mammifères euthériens. Liste critique et essai sur l'évolution chromosomique. *Archiv Julius Klaus-Stift*, **33**, 253–97.

Mayer, W. V. (1952). The hair of mammals with keys to the dorsal gland hairs of California mammals, *American Midland Naturalist*, **48**, 480–512.

Meyer, G. E. (1978). Hyracoidea. In *Evolution of African mammals* (ed. V. J. Maglio and H. B. S. Cooke), pp. 284–314, Harvard University Press.

Nadler, C. F., Lyapunova, E. A., Hoffman, R. S., Vorontsov, N. N., and Malygina, N. A. (1975). Chromosomal evolution in holarctic ground squirrels (*Spermophilus*). I Giemsa band homologies in *Spermophilus columbianus* and *S. undulatus*. *Zeitschrift fur Säugetierkunde*, **40**, 1–7.

Napier, J. (1971). Paleoecology and catarrhine evolution. In *Old World Monkeys* (ed. J. R. Napier and P. H. Napier), pp. 55–95. Academic Press, London.

Nelson, G. (1979). Cladistic analysis and synthesis: principles and definitions, with a historical note on Adanson's *Familles des Blantes* (1763–1764), *Systematic Zoology*, **28**, 1–21.

Nelson, G. and Platnick, N. (1981). *Systematics and biogeography*, Columbia University Press, New York.

Parent, J. P. (1976*a*). La région auditive des rongeurs sciurognathes. Caractères anatomiques fondamentaux. *Comptes Rendus de l'Académie des Sciences, Paris*. **282**, 2183–5.

Parent, J. P. (1976*b*). Disposition fondamentale et variabilité de la région auditive des rongeurs hystricomorphes. *Comptes Rendus de l'Académie des Sciences, Paris*. **283**, 243–5.

Parent, J. P. (1980). Recherches sur l'oreille moyenne des rongeurs actuels et fossiles. Anatomie, valeur systématique. *Mémoires et Travaux de l'E.P.H.E., Institut de Montpellier*. **11**, 1–286.

Patterson, B. and Wood, A. E. (1982). Rodents from the Deseadan Oligocene of Bolivia and the relationships of the Caviomorpha. *Bulletin of the Museum of Comparative Zoology, Harvard*. **149**, 371–453.

Patton, J. L. and Gardner, A. R. (1972). Notes on the systematics of *Proechimys* (Rodentia Echimyidae), with emphasis on Peruvian forms. *Occasional papers, Museum of Zoology, Louisiana State University*. No. 44, 1–30.

Pilleri, G. (1960). Comparative anatomical investigations on the central nervous system of rodents and relationships between brain form and taxonomy. *Revue suisse de Zoologia*, **67**, 373–86.

Pocock, R. I. (1922). On the external characters of some hystricomorph rodents. *Proceedings of the Zoological Society of London*, **?**, 365–427.

Quentin, J. C. (1973*a*). Affinités entre les oxyures parasites des rongeurs hystricidés éréthizontidés et dinomyidés. Intérêt paléobiogeographique. *Comptes Rendus de l'Académie des Sciences, Paris*, **276**, 2015–17.

Quentin, J. C. (1973). Les oxyures de rongeurs. *Bulletin du Musée National d'Histoire Naturelle de Paris*, **167**, 1046-96.

Raza, S. M., Barry, J. C., Pilbeam, D., Rose, M. D., Shah, S. M. I., and Ward, S. (1983). New hominoid primates from the middle Miocene Chingi formation, Potwar plateau, Pakistan, *Nature*, **306**, 52–64.

Ross, H. H. (1974). *Biological systematics*. Addison-Wesley, Reading, Massachusetts.

Savage, D. E. and Russell, D. E. (1983). *Mammalian paleofaunas of the world*. Addison-Wesley, Reading, Massachusetts.

Savage, R. J. G. (1978). Carnivora. In *Evolution of African mammals*, (ed. V. J. Maglio and H. B. S. Cooke), pp. 249–67, Harvard University Press.

Shevyreva, N. S. (1972). New rodents from the Paleogene of Mongolia and Kazakhstan. *Akademia Nauk SSSR, Paleontological Journal*, **3**, 136–45.

Simons, E. L. (1968). African Oligocene mammals: Introduction, history of study, and faunal succession. In Early Cenozoic mammalian faunas, Fayum province, Egypt. *Peabody Museum of Natural History Bulletin*, **28**, 1–91.

Simpson, G. G. (1978). Early mammalian South America. Facts, controversy and a mystery. *Proceedings of the American Philosophical Society*, **122**, 318–28.

Simpson, G. G. (1980). *Splendid isolation*, Yale University Press, New Haven.

Singh, G. (1988). History of aridland vegetation and

climate: a global perspective. *Biological Review*, **63**, 159–95.

Sudre, J. (1979). Nouveaux mammifères éocènes du Sahara occidental. *Palaeovertebrata*, **9**, 83–115.

Stromer, E. (1923). Bemerküngen über ersten Landwirbeltierreste aus dem Tertiär Deutsch Südwestafrikas. *Paleontologische Zeitschrift*, **5**, 226–8.

Tarling, D. H. (1980). The geologic evolution of South America with special reference to the last 200 million years. In *Evolutionary biology of the New World monkeys and continental drift* (ed. R. L. Ciochon and A. B. Chiarelli), pp. 1–41, Plenum Press, New York.

Thenius, E. (1976). Zur Heikunft der Wasserschweine (Hydrochoerus, Rodentia, Mammalia), *Zeitschrift Säugetierkunde*, **41**, 250–2.

Thomas, O. (1896). On the genera of the rodents: An attempt to bring up to date the current arrangement of the order. *Proceedings of the Zoological Society of London*, 1012–28.

Toepfer, K. (1897). Die Morphologie des Magens der Rodentia. *Morphologische Jahrbuch*, **17**, 380–407.

Tullberg, T. (1899). *Ueber das system der Nagethiere. Eine phylogenetische studie*. Nova Acta Regiae Societatis Scientificae Upsala, Ser. 3, **18**, 1–514.

Vianey-Liaud, M. (1985). Possible evolutionary relationships among Eocene and lower Oligocene rodents of Asia, Europe and North America. In *Evolutionary relationships among rodents. A multidisciplinary analysis* (ed. W. P. Luckett and J. L. Hartenberger), pp. 277-309, Plenum Press, New York.

Webb, S. D. (1978). A history of savanna vertebrates in the New World. Part II: South America and the great interchange. *Annual Review of Ecological Systems*, **9**, 393–426.

Weir, B. J. (1974). Reproductive characteristics of hystricomorph rodents. In *The biology of Hystricomorph rodents* (ed. I. W. Rowlands and B. J. Weir) pp. 265–301, Academic Press, London.

White, B. N. (1986). The isthmian link, antitropicality and American biogeography: distributional history of the Atheronopsinae (Pisces: Atherinidae), *Systematic Zoology*, **35**, 176–94.

Wiley, E. O. (1981). *Phylogenetics*. John Wiley, New York.

Wood, A. E. (1955). A revised classification of the rodents. *Journal of Mammalogy*, **36**, 165–87.

Wood, A. E. (1965). Grades and clades among rodents. *Evolution*, **19**, 115–30.

Wood, A. E. (1968). The African Oligocene rodentia. In Early Cenozoic mammalian faunas, Fayum province, Egypt. *Bulletin of the Peabody Museum of Natural History*, **28**, 23–105.

Wood, A. E. (1972). An Eocene hystricognathous rodent from Texas: its significance in interpretations of continental drift. *Science*, **175**, 1250–1.

Wood, A. E. (1974). The evolution of the Old World and New World hystricomorphs. In *The biology of Hystricomorph rodents* (ed. I. W. Rowlands and B. J. Weir), pp. 265–301, Academic Press, London.

Wood, A. E. (1977). The evolution of the Rodent family Ctenodactylidae, *Alexandrovich Orlov Memorial Number, Journal of the Paleontological Society of India*, **20**, 120–37.

Wood, A. E. (1980). The origin of the Caviomorph rodents from a source in Middle America: a clue to the area of origin of the platyrrhine primates. In *Evolutionary biology of the New World monkeys and continental drift* (ed. R. L. Ciochon and A. B. Chiarelli), pp. 79-91, Plenum Press, New York.

Wood, A. E. (1984). Hystricognathy in the North American Oligocene rodent *Cylindrodon* and the origin of the Caviomorpha. In Papers in vertebrate paleontology Honoring Robert Wilson (ed. R. M. Mengel), pp. 151–60. *Carnegie Museum of Natural History, Special Publication* 2.

Wood, A. E. (1985). The relationships, origin and dispersal of the hystricognathous rodents. In *Evolutionary relationships among rodents. A multidisciplinary analysis* (ed. W. P. Luckett and J. L. Hartenberger), pp. 475–513, Plenum Press, New York.

Wood, A. E. (1985). Northern waif primates and rodents. In *The great american biotic interchange* (ed. F. G. Stehli and S. D. Webb), pp. 267–82, Plenum Press, New York.

Wood, A. E. and Patterson, B. (1959). The rodents of the Deseadan Oligocene of Patagonia and the beginnings of South American rodent evolution. *Bulletin of Museum of Comparative Zoology, Harvard*, 281–428.

Wood, A. E. and Patterson, B. (1970). Relationships among hystricognathous and hystricomorphous rodents. *Mammalia*, **34**, 628–39.

Woods, C. A. (1972). Comparative myology of jaw, hyoid and pectoral appendicular regions of New and Old World hystricomorph rodents. *Bulletin of the American Museum of Natural History*, **147**, 115–98.

Woods, C. A. (1975). The hyoid, laryngeal and pharyngeal regions of Bathyergids and other selected rodents. *Journal of Morphology*, **147**, 229–50.

Woods, C. A. and Hermanson, J. W. (1985). Myology of hystricognath rodents: An analysis of form, function, and phylogeny. In *Evolutionary relationships* *among rodents. A multidisciplinary analysis* (ed. W. P. Luckett and J. L. Hartenberger), pp. 515-48, Plenum Press, New York.

10 CONCLUSIONS
René Lavocat

As discussed by A. E. Wood in the introduction to this book, the question of the Africa–South America connection, evidenced by more or less apparent parallels in the flora and fauna of both continents, took on an entirely new dimension when plate tectonics demonstrated that today's distinct continents were once united in a single mass, before being slowly separated by continental drift. For many years, the apparent similarities between the various faunal and floral elements were used as key arguments supporting, among other things, the existence of what was called the Gondwana continent. Now that we have indisputable geological proof of this phenomenon, it is probably less important to provide biological evidence for it. In fact, we should use our knowledge of the existence of the continent to understand further the biological relationships. Nevertheless, plate tectonics, particularly during transition stages, may not always reveal the precise timing or extent of continental movements. For example, the presence of a shallow epicontinental sea, which would not be revealed by plate tectonics, could have restricted migrations routes and hampered the migration of some animals. Clearly, such a sea would have been as difficult to cross as a deep ocean for small terrestrial vertebrates, but not for huge amphibian hadrosaurians.

Knowledge of the fauna and flora is thus of special interest when it is related to questions of timing and distribution. When were the direct land routes between the two continental masses of Africa and South America cut? As the continents drifted apart, when did migration become improbable? When did it become impossible? How does the continents' history explai the relationships and present distribution of plants and animals?

The problem is further complicated by the role played by climate in the dispersal and distribution of living things. On one hand, the position of continental masses changed relative to the Equator and to the climatic zones existing at that time. On the other hand, climate itself has varied greatly through time, with wide extensions of the tropical and warm temperate zones as far as the Arctic circle (Ellesmere), followed by worldwide cooling. The various positions of the land masses have influenced significantly these climatic variations. Sea-level itself, as a consequence of both tectonic events and climatic changes, has at times affected the climate at least locally, and has physically affected possible land routes through its variations of more than 100 metres.

The present populations of the Earth are thus largely the result of these two main factors, local tectonic movements and climatic variation, and of their complex effects. We must bear in mind that a climatic barrier, for example a desert zone, can be even more effective than an oceanic barrier in preventing migration. A theoretically open land route may be useless under these circumstances. Changes in the climate can thus sever the biological unity of a continental land mass. As pointed out by Amedegnato, knowledge of the existence of a united continent can be a dangerous basis for a proper appraisal of the facts. Continental unity may sometimes result in biological homogeneity, but this does not necessarily occur. Even if it appears likely that a group of animals arose on the united continent because of the group's age, any firm conclusion is impossible without fossil evidence. For example, the history of palaeontology has demonstrated how it can be unwise to assume that the absence of a fossil form, in an inadequately explored region, is meaningful. For more than a century, marsupials were believed to have never occurred in Africa!

Many groups appeared later, shortly before or

after the separation of the continents. As the plates moved apart, communication between the continents became increasingly difficult. Communication processes other than land connections must therefore have been important for the dispersal of these late-arising groups. This book gives excellent examples of some of the various ways in which the problems of the Africa–South America connection can be expressed.

Parrish offers a solid *mise-au-point* of the relevance of plate tectonics for understanding the timing and initial events leading to the separation of the two continents, and the eventual difficulties of migration after the separation. She also gives important information about the problem of possible migration occurring between both Americas while Central America and the Antilles were forming.

My personal feeling is that the existence of islands of unknown extent in the Atlantic, mainly in the Eocene and Oligocene, has a very important bearing on the likelihood of ocean crossing if it implies a seaward extension of the continental coast. Such an extension would have shortened the oceanic distance to be travelled by migrating animals.[1] This hypothetical high-level crest is also important if a concomitant lowering of the sea-level resulted in an uninterrupted land freeway between the two continents; however, this seems unlikely. In my opinion, given the comparative sizes of the Brazilian coast and of the possible islands in the Atlantic, it seems more likely that a raft would have landed on the continental coast than on the islands. Moreover, reaching an island while crossing the Atlantic results in a real and difficult problem, i.e. sailing away from it. We cannot expect that an island of limited size should be able to produce or even less to consign a raft to

the open sea with the same efficiency as a powerful African equatorial river. However, such scattered islands could well have acted as relay stations or resting areas for birds and flying insects, or eventually amphibious animals. (Amedegnato pointed out to me that grasshoppers are able to land and float on the sea and, after resting, resume their flight. The presence of islands may thus be unimportant for some groups.) It seems to me rather unlikely that such islands could really help the migration of passively transported animals; much more likely, such animals would remain captive on the islands, but I welcome proof to the contrary.

Palaeontological evidence for land connections between North and South America in the Tertiary is much weaker than geological evidence. If such land connections did exist, their effects on faunal distribution should be as clearly seen as those in the 'Great Exchange', but they are not. Some obstacles, which plate tectonics cannot detect, must have been in place for most of the Tertiary, and certainly during the Eocene.

Plants, particularly angiosperms, allow us to examine two important aspects of the problems evoked above. Melville (1973) believed that several leaf samples of the Jurassic, found in England, were angiosperms, but he acknowledged the fact that Hughes (1961) held different views. Van Campo (1978) wrote:

L'apparente diversité des mégafossiles d'Angiospermes du Crétacé inférieur, principalement des feuilles, a longtemps été considérée comme la preuve que les Angiospermes avaient déjà un long passé d'évolution antérieur au Crétacé; récemment un réexamen critique à la fois sur le plan morphologique et sur le plan stratigraphique des feuilles d'Angiospermes du Crétacé a montré que l'ensemble des fossiles corroborait le concept palynologique de la radiation adaptative des Angiospermes pendant le Crétacé.

One can thus reasonably suppose that the onset of angiosperm radiation occurred in the Lower Cretaceous. Of course, the Cretaceous lasted for about 70 million years, longer than the Tertiary, but when the Albian began, evidence from plate tectonics suggests that communication between

---

[1] This may have influenced the exact time when direct connections between Africa and South America were no longer possible. Herngreen (1974) insists that the comparison between the evolutionary processes of microfloras of South America and Western Africa proves that up to the Albian, and perhaps even the Cenomanian, a direct connection by some land route between the two continents still existed. See also information about the African microflora in Jardine *et al.* (1974).

Africa and South America may have no longer existed. Some palaeontologists (e.g. Buffetaut, this volume) doubt this, so conclusions should be drawn carefully. However, even if angiosperms did arise in the austral countries, the lack of a good fossil record precludes confirmation that a common population originated on the united continents. Thus, if some close relationships exist between the plants of the Africa and South America, the similarities will have to be ascribed, due to a lack of other evidence, to methods of dispersal during the Tertiary. In light of the presently known facts, any intepretation will have to remain more or less provisional.

The comparison between the Aroaceae, discussed in this book by Mayo, and the Leguminosae, discussed by Raven and Polhill (1981) in another symposium, illustrates two different approaches to the same lack of fossils. Mayo gives a description of the territorial interrelationships found within the Aroceae, which is enhanced by our knowledge of the Tertiary (in the northern hemisphere) and of present distributions. He concludes that 'what emerges is a pattern of relationships that clearly links Africa and South America individually to Asia, but only rather tenuously to each other'. His paper demonstrates affinities between African and South-east Asian plants, and between South and North American plants, the latter being close to plants of the other northern continents. If there were a population common to Africa and South America in Cretaceous time, it seems clear that, in this particular situation, the resulting separate populations were altered by severe extinctions and subsequent migrations and substitutions.

By contrast, Raven and Polhill (1981) in their paper *Biogeography of the Leguminosae* emphasize the richness of the related populations of this group common to Africa and South America. In both continents, they find species of common and allied genera, whereas few are found in Eurasia. The authors conclude that without doubt the legumes of tropical North America are essentially derived from those of South America, while those of the temperate regions of North America are

mainly Eurasian in origin. They also think that South America received its legumes from Africa which, they think, was widely separated from South America at the time of differentiation of the main groups of legumes.

The contrast between these two groups of angiosperms is striking, likely as a result of the great differences in the capacities for dispersal of each group. The seeds of Aroaceae do not seem able to survive in salt water while, according to Raven and Polhill (1981), this does not pose a problem for the Leguminosae. Their seeds have apparently had not difficulty in reaching Hawaii. Indeed, even now the Brazilian coast is nearer to the African coast than Hawaii is to any continent. Moreover, the slow rate of drift after the Cretaceous separation would have maintained the coasts of the two austral continents in close proximity for a very long time. At the same time, the equatorial current and the trade-winds could have favoured east-west travel. The authors refer to rodent and primate rafting; I think that, for example from the Hawaiian evidence, they strongly reinforce our common argument. It is also clear that a comparison of the two plant families demonstrates how much the close relationships depend on the possibility of direct, in this case transoceanic, migrations.

In her paper on acridians (this book), Amedegnato shows that the arrival of *Schistocerca* in South America proves, with absolute certainty, the efficiency of westward winds as aids to dispersal; the Leguminosae testify to the efficiency of oceanic equatorial currents, also in a westward direction. It is on these two factors that rodent and primate migration relies.

Amedegnato's Acridoidea present us with the same problem as the Aroaceae, i.e. a lack of fossils. Indeed, whatever our convictions maybe about the origin of this insect group, or whether its ancestors inhabited the Gondwana continent, the complete lack of known fossils in South America and the young age of fossils found in other continents do not allow us to establish clearly the early relationships between Africa and South America. The present relationships may more likely be the result of the great radiation of the acridians in the northern continents, up to the polar regions, helped by favourable climates at the beginning of the Tertiary. Thereafter, this range extension was followed by migrations of colonization or recolonization towards Africa and South America. If the southern continents sheltered true ancestors of the acridians, why we find no evidence for this in more recent populations remains a mystery. It is impossible to reconstruct the past by examining present-day geography. However, it is very interesting to note that the single element which suggests a possible relationship between the two great southern continents is the demonstration of the extraordinary wind-aided migratory ability of *Schistocerca*. As Amedegnato suggests, there is no reason to think that the recently observed successful crossing towards South America should be an event unique in history; the ability for flight of this genus has been demonstrated, and flight ability may often be enhanced by good winds. Of course, the ability to migrate does not necessarily reflect the ability to colonize, and Amedegnato emphasizes that the sucessful ocean crossing does not prove successful colonization, nor the effective establishment of some of these migrants, to the extent that the forms settled in South America can be related to others of North America. Not being a specialist on Orthoptera, it is with the greatest humility that I ask this question: is it not meaningful that these North American forms are found in Cuba and Florida, the south-easternmost parts of the North American continent? If ancient fliers could have reached North America, surely we should expect to find them in these regions. The specialists will judge.

In 1973, Romer was deploring the rarity of useful vertebrate fossils for studying the problem of relationships between Africa and South America during the Cretaceous. In the same symposium, Keast (1973) showed from a wide biological survey, what could be deduced from the comparison of the populations of the two continents, while Cracraft (1973) gave an account of the rare palaeontological data then available. Since that time, the contributions of Buffetaut and Rage have led to an important updating of information for several reptilian and amphibian groups. Interestingly, these animals were present around the critical tectonic period of the Cretaceous; their work thus helps to determine when the land connections were severed. Of equal interest is the authors' belief that some communications for several animals still remained after the break, implying that there were some migrations over the newly formed Atlantic. Since the animals discussed could not tolerate salt water, they must have left the African coast on rafts.

The Eocene locality at Messel, near Darmstadt, in Germany (Schaal and Ziegler 1988) is without doubt a unique scientific wonder in the world of palaeontology. It is a natural museum of the most exceptional interest, gathering an extraordinary number of rare documents beautifully preserved, a scientific treasure worthy of the most careful attention. Most remarkable is the presence in Eocene (Lutetian) time of several vertebrates (a mammal, two birds, and a reptile) showing strong South American affinities. To find tropical animals in Germany's Eocene past is interesting, but not really surprising. We know that the global climate was warm even in very northern latitudes, as proven by the plants of northern islands and palaeontological discoveries at Ellesmere Island on the Arctic circle (West and Dawson 1978). What seems more surprising is that there are good reasons to believe that these animals originated in South America, and that the most probable migration route was via Africa. There were definite faunal relationships between Africa and South America at the time of the Cretaceous (Buffetau and Rage, this volume). (Crochet (1948), a specialist of marsupials, described one of these mammals in the Eocene of Africa.)

In 1969, Lavocat reached the conclusion that the strong anatomical relationship between the Phiomorpha and Caviomorpha could be explained by a direct migration, in the Palaeogene, by rafts across a narrow Atlantic. Hoffstetter immediately accepted this hypothesis; the strong arguments provided by the rodents made it likely that the idea would also apply to primates. For this reason, Lavocat was asked to put forward the arguments resulting from the study of rodents to the Primatological Congress of Bangalore (Lavocat 1980). For nearly 20 years, Lavocat and Hoffstetter both maintained the idea of a common scenario for the distribution of hystricognath rodents and primates, while their opinion was strongly debated by specialists. Recently, palaeontological discoveries have given new evidence relating specifically to primates.

Aiello (this book) writes a very clear statement of the different hypotheses of dispersal proposed for primates and of the supporting observations, and provides an objective evaluation of each argument. Not being a primate specialist, I cannot judge the relative importance of these arguments. However, I see with great interest that the hypothesis put forward by Hoffstetter is considered a most probable model. Her conclusions may not be universally accepted, and may be viewed by some as controversial, but such is the fate of many scientific opinions. I should like to quote from a first draft of her contribution written some years ago, but not published. At that time, Aiello wrote:

On the grounds that the most probable model would require the least number of waif dispersal events, preference can be given to a southern continents origin of the anthropoid primates (and primates in general), and an African origin of the platyrrhine monkeys.

and in later conclusions, she remarked cautiously:

In spite of the possible preference for the African centred model on the grounds of parsimony of dispersal events, the obvious weak point of this model is the lack of fossil evidence in the southern continents to provide it with conclusive palaeontological support.

Both her work and her judgement have been validated by the recent discovery of *Altiatlasius koulchii*, the first Palaeocene primate found in Africa (Sigé *et al.* 1990). It is not necessary to dwell on the significance of this find for Aiello's conclusions, or for our understanding the history of primates. We know that H. Thomas and coworkers' discoveries in the Oman sultanate in recent years have demonstrated the great exten-

sion of various Primates over the African plate in the Earliest Oligocene or even the Uppermost Eocene. Furthermore, some very recent publications about Northern Africa definitely show the great antiquity of the diverse population of Primates in Africa.

It is reasonable to state that knowledge on this subject has been completely transformed, so as to favour strongly the African theory.

Let us now come to the rodents and to the contribution made by Wilma George to the study of this group of animals. She summarizes the essential points of the friendly and vigorous discussion between Wood, supporter of the North American origin of the South American Caviomorpha, and Lavocat, supporter of their African origin. This was a useful and fruitful discussion, which offered us the opportunity to examine all aspects of the problem and to weigh the pros and cons, as was our scientific duty. All the details can be found in the bibliography, and I shall not repeat what has already been explained. Wilma George, who contributed in many ways to the study of the systematic affinities among rodents, has drawn up an impressive list of characters which have been studied comparatively. One wonders if it is possible to touch many more fields. She extracts from this mass of information valuable results which, I think, can be taken as conclusive evidence for strong affinities between the Phiomorpha and the Caviomorpha. Clearly, I agree with her opinion that relationships covering several thousands of kilometres of land routes and crossing many degrees of latitude and longitude are not very satisfying.

To her analysis, I will add some recent results and opinions. The preliminary study by Lavocat (1988) on the Fort Ternan (Kenya Miocene) Bathyergidae *Richardus excavans* was followed, in 1989, by a detailed anatomical study which confirmed the hypothesis of a reduction of the infraorbital foramen, previously of phiomorph type. *Richardus* cannot derive from *Tsaganomys*, whose ear region does not support such a relationship (Parent and Lavocat, unpublished.)

Wood favoured a North American group of the

Lower Eocene as the ancestors of the Caviomorpha. Among the former is *Reithroparamys delicatissimus*, an early representative of the Franimorpha, based on the assumption that these rodents were incipiently hystricognathous. Several palaeontologists (Dawson, Korth) do not agree with this interpretation. Moreover, Jin (1990) has shown that, at least in *Reithroparamys*, the detailed anatomy of the ear precludes any eventual relationship with the Caviomorpha.

More recently, in a detailed study of the enamel structure of the incisors of the Hystricognathi and North American ancient rodents, Martin (1990, 1992) showed that no North American rodent proposed as ancestral to the Caviomorpha shows a tendency toward the typical caviomorph multiserial enamel structure. On the other hand, the structures of the Phiomorpha and Caviomorpha are perfectly coherent. As a consequence, Martin concluded that his results add weight to other arguments supporting the African origin of the Caviomorpha from a Phiomorpha ancestor, via a raft crossing of the Atlantic.

On the Indian subcontinent of the Old World, a group of peculiar rodents, the Chapattimyidae, existed in the Middle Eocene. Hussain *et al.* (1978) suggested that these were the ancestors of the Phiomorpha and Hystricidae on one side and of the Caviomorpha on the other side, via a great transcontinental migration. But in 1982 they wrote:

Deriving the Phiomorpha from the Chapattimyidae still seems quite possible and even probable, but to see this family as the source of the Hystricidae and Caviomorpha is, as Wilson correctly analysed (unpublished manuscript 1981), nothing but a tantalizing hypothesis.

Personally, I cannot believe that the Chapattimyidae could be the ancestors of the Phiomorpha. One important reason is that they show a strong metaconule while this tubercle is absent, or greatly reduced, in the Phiomorpha. Jaeger *et al.* (1985) placed the Chapattimyidae as the sister group of the Phiomorpha. From their explanation, this may be possible. The discovery in Spain of Middle Eocene rodents morphologically identical to the Chapattimyidae (Pelaez-Campomanes *et al.* 1989) raises additional questions. It is true that in Europe, *Adelomys cartieri*, with a strong metaconule, and *Adelomys vaillanti*, with a much less pronounced one, are described byStehlin and Schaub 1951. Clearly, the Theridomyidae are derived from this second morphological type, and I persist in my opinion, expressed a few years ago, that some rodents close to the ancestors of the Theridomyidae, which migrated perhaps as early as the Lower Eocene in Africa, could be the ancestors of the Phiomorpha. In the Eocene of Nementcha (Algeria), anomalurids are found, in addition to *Protophiomys*. Until now, no better place for the origin of the anomalurids has been found than Europe, and it is suspected that they could be allied to the Theridomorpha. Of course, the ancestors of the Theridomorpha were certainly sciurognathous, but if the Chapattimyidae are ctenodactylids, they are then also sciurognathous. It is reasonable to describe sciurognathy as a primitive structure and hystricognathy as an evolved, or derived, character.

Wood (this volume) quotes a paper by Webb (1978) which states that a raft would have had to start from the region of the Congo's mouth, i.e. a very long way from Brazil. But, have we any evidence that in the Palaeogene the Equator crossed Africa at the same latitude as it does now? There is reason to believe that Africa, still rotating and moving northward, was then in a more southern position than it is now. That shows, I think, that the equatorial rains probably fell more on the Fouta Djalon than on the Congo basin. Henced a raft could have started just as easily from the most western coast of Africa (i.e. 2000 kilometres closer to Brazil), helped by a combination of westward winds and currents. I am not as pessimistic as Simons about the fate of the small travellers; they could have taken advantage of the humidity and rains of the equatorial regions, and even found green leaves and fruit to eat for some weeks on partly preserved trees. A raft with trees is not a bare log!

Conclusion to the conclusion

Thanks to the authors of the many contributions in this book, we have been able to examine the behaviour of various groups in the colonization of the two great southern continents. From that survey, the following results emerge. Some groups were physically unable to cross the Atlantic barrier. The colonization of Africa and South America by these groups, in the Tertiary, was possible only via the northern continents over a direct land connection with either of the southern ones. In these cases, the systematic relationships between the two sides of the Atlantic remain clearly distant. Apart from the primates and rodents, plants of the Leguminosae from both southern continents show close affinities. In this case, the analysis shows that South America was not colonized from North America, but that the plants travelled from Africa across the sea. These were plants whose present-day relatives have seeds that can float as far as Hawaii. The contrast between the two situations is clear: close taxonomic relationships may be correlated with direct transoceanic connections. Thus, the facts agree with theoretical evidence.

Having rejected *Reithroparamys* as an ancestor for the Caviomorpha, we cannot conclude definitely that potential ancestors for caviomorph rodents cannot be found in North America. However, two lines of evidence point in a different direction. First, there are such overwhelming data showing affinities between the African Phiomorpha and the South American Caviomorpha that one cannot deny their close systematic proximity. Second, rodents clearly related to the Fayum Phiomorpha are known in the African Eocene, and genuine rodents have been present in northern Africa[1] since the early Eocene.

---

[1] Comments by Dr Hartenberger: 'After reading a first draft of this paper by Dr Lavocat, I confirm the antiquity of the first known fossil of rodents in African Early Eocene. The locality of Chambi in Tunisia (Hartenberger *et al.* 1985) yielded, between other mammals, very few rodent teeth. There is no doubt about the Early Eocene age of this locality, and, even with this small sample, two families are present: one is the typical African anomalurid, the other could be an ischyromid.'

Helped by the conclusions drawn by comparing the different means of colonization, we can reasonably affirm that there must have been a direct connection through the Atlantic. So, if we postulate a North American ancestor, a first crossing of the sea from north to south must have been necessary, followed by successful colonization. Increases in the population must have occurred to permit a second migration, this time towards Africa. The west–east current which would allow this crossing was a long way south, and the shape of the coastline made the crossing much longer this way than east–west via the equatorial current. The supporters of the hypothesis of a North American rodent origin have several times volunteered that this east–west sea crossing would have been impossible. But the close affinities between the two faunas require that the crossing must have been made in one direction or the other. Is it too much to prefer the easiest route?

Moreover, the more difficult west–east crossing had to be accomplished early enough to explain the presence of Phiomorpha in northern Africa in the Eocene. Remember that the starting place must have been somewhere in southern Argentina. Pascual (*in litt.*) collected a great many fossils of small vertebrates by wash-sampling Eocene material from this area—but found no rodents. That should give us food for thought, and not just about rodents!

REFERENCES

Cracraft, J. (1973). Vertebrate evolution and biogeography in the Old World tropics: implications of continental drift and palaeoclimatology. In *Implications of continental drift to the earth sciences.* Vol. 1., (ed. D. H. Tarling and S. K. Runcorn), pp. 373–93. Academic Press, London.

Crochet, J. T. (1984). *Peratherium mahboubii* nov. gen., nov. sp., marsupial de l'Eocène inférieur d'El Kohol (Sud-oranais, Algérie). *Annales de Paléontologie*, **70**, 275–94.

de Bruijn, H., Hussain, S. T., and Leinders, J. J. M. (1982). On some early Eocene rodent remains from Barbara Banda, Kohat, Pakistan, and the early history of the Order Rodentia. *Proceedings of the Koninklijke Nederlandse Akademie van Weten-schappen*, Series B, **85**, 249–58.

Herngreen, G. F. W., (1974). Middle Cretaceous palynomorphs from northeastern Brazil. Results of a palynological study of some boreholes and comparison with Africa and the middle East. In *Palynologie et dérive des continents*. Sciences Géologiques, Bulletin 27, pp. 101–17, Institut de Géologie, Strasbourg.

Hughes, N. F. (1961). Fossil evidence and Angiosperm ancestry. *Science Progress*, **49**, 84–102.

Hussain, S. T., de Bruijn, H., and Leinders, J. J. M. (1978). Middle Eocene rodents from the Kala Chitta Range (Punjab, Pakistan). *Proceedings of the Koninklijke Nederlandse Akademie van Wetenschappen*, series B, **81**, 74–112.

Jaeger, J. J., Denys, C., and Coiffait, R. (1985). New Phiomorpha and Anomaluridae from the late Eocene of north-west Africa: phylogenetic implications. In *Evolutionary relationships among rodents. A multidisciplinary analysis* (ed. W. P. Luckett and J. L. Hartenberger), pp. 567–88. Plenum Press, New York.

Jardiné, S., Kieser, C. and Reyre, Y. (1974). L'individualisation progressive du continent africain vue à travers les données palynologiques de l'Ere secondaire. In *Palynologie et dérive des continents*. Sciences Géologiques. Bulletin, **27**, pp. 69-85, Institut de Géologie, Strasbourg.

Jin, M. (1990). The auditory region of *Reithroparamys delicatissimus* (Mammalia, Rodentia) and its systematic implications. *American Museum Novitates*, **2972**, 1–35.

Keast, A. (1973). Contemporary biotas and the separation sequence of the southern continents. In *Implications of continental drift to the earth sciences*, Vol. 1., (ed. D. H. Tarling and S. K. Runcorn), pp. 309–43. Academic Press, London.

Lavocat, R. (1969). La systématique des Rongeurs hystricomorphes et la dérive des continents. *Comptes Rendus de l'Académie des Sciences*, Paris, **269**, 1496–7.

Lavocat, R. (1980). The implications of Rodent paleontology and biogeography to the geographical sources and origin of the Platyrrhine Primates. In *Evolutionary Biology of the New World Monkeys and Continental Drift* (ed. R. L. Ciochon and A. B. Chiarelli), pp. 93–102.

Lavocat, R. (1988). Un rongeur bathyergidé nouveau remarquable du Miocène de Fort Ternan. *Comptes Rendus de l'Académie des Sciences*, Paris, **306**, 1301–14.

Lavocat, R. (1989). Ostéologie de la tête de *Richardus excavans* Lavocat, 1988. *Palaeovertebrata*, **19**, 73–80.

Hartenberger, J. L., Martinez, C., and Ben Said, A. (1985). Découverte de mammifères d'âge Eocène inférieur en Tunisie centrale. *Comptes rendus de l'Académie des Science*, Paris, **301**, ser. II, 649–52.

Martin, T. (1990). Origin of the caviomorphs: evidence from incisor enamel. *Journal of Invertebrate Paleontology*, **10**(3) (Abst), 34A.

Martin, T. (1992). Enamel microstructure in the incisors of Old and New World hystricognath rodents. *Palaeovertebrata* (in German) (In press).

Melville, R. (1973). Continental drift and plant distribution. In *Implications of continental drift to the earth sciences*, Vol. 1., (ed. D. H. Tarling and S. K. Runcorn), pp. 439–46. Academic Press, London.

Pelaez-Campomanes, P., De la Pena, A., and Lopez Martinez, N. (1989). Primeras faunas de micro-mamiferos del Paleogeno de la cuenca del Duero. In *Paleogeografia de la Meseta norte durante el Terciario*. Studia geologica salmanticensia, Vol. sp. **5**, (ed. C. J. Dabrio), pp. 135–57. Ediciones Universidad de Salamanca.

Raven, P. H. and Polhill, R. M. (1981). Biogeography of the Leguminosae. In *Advances in Legume systematics* (ed. R. M. Polhill and P. H. Raven), pp. 27–34.

Romer, A. S. (1973). Vertebrates and continental connections: an introduction. In *Implications of continental drift to the earth sciences*, Vol. 1 (ed. D. H. Tarling and S. K. Runcorn), pp. 345–49. Academic Press, London.

Schaal, S. and Ziegler, W. (ed.) (1988). *Messel: ein Schaufenster in die Geschichte der Erde und des Lebens*. (Ed. S. Schaal and W. Ziegler), pp. 1–314. Kramer, Frankfurt am Main.

Sigé, B., Jaeger, J. J., Sudre, J., Vianey-Liaud, M. (1990). *Altiatlasius koulchii* n. gen. et sp., primate omomyidé du Paléocène supérieur du Maroc, et les origines des euprimates. *Paleontographica Abt. A.*, **214**, 31–56.

Stehlin, H. G. and Schaub, S. (1951). Die Trigonodontie der simplicidentaten Nager. *Schweizerische Palaeontologische Abhandlungen*, **67**, 1–385.

Van Campo, M. (1978). Phylogénie des Angiospermes, approche palynologique. In *Aspects modernes des recherches sur l'évolution*, Vol.1. Mémoires et Travaux E.P.H.E. Institut de Montpellier, **4**, pp. 73–87.

Webb, S. D. (1978). A history of savanna vertebrates in the New World. Part II: South America and the great interchange. *Annual Review of Ecology and Systematics*, **9**, 393–426.

West, R. M. and Dawson, M. R. (1978). Vertebrate paleontology and the Cenozoic history of the North Atlantic region. Polarforschung, **48**, 103–19.

Wood, A. E. See the references given by W. George (this volume).

APPENDIX: THE HISTORY OF THE PROBLEM

Wilma George

When Wilma George died, after a long illness, her chapter on the rodents (p. 119 of this volume) was available but no trace could then be found of the chapter on the history of the Africa–South America problem that she had planned, apart from a brief outline. In order to complete the scheme for the book as it was originally envisaged by Wilma George, Professor A. E. Wood, who has a long and deep interest in this particular problem, was asked to write the historical account that appears as Chapter 1. It was only after he had delivered his manuscript that, by a stroke of good fortune, Wilma George's manuscript was discovered. The text appears to be in its final form, and it is reproduced below, together with the illustrations that Wilma George had assembled for it.

The reader will, we think, find special interest in this account of earlier thought on the subject, for Wilma George had access to, and was familiar with, a wide range of literature that would not readily be available elsewhere.

R.L.

The discovery of America at the end of the fifteenth century revealed not only new races of men and the potential for gold but also strange new plants and animals. So exciting were the new discoveries that men, plants, and animals were rapidly incorporated into the maps of the period which were thus the first biogeographical maps (George 1969). The travellers tales, exhibits, and the maps stimulated speculation on organic distribution. The long known differences between Africa and Europe could be explained away by differences in climate, but the striking faunal and floral differences between Africa and South America—countries with similar climates—seemed to require some other explanation.

As the opossums, capuchin monkeys, macaws, and cactus were shipped back to Europe it became clear that 'the infinite variety of sylvan animals', as Amerigo Vespucci had remarked in 1500, were 'not like those of our regions'. And it was difficult to imagine how so many 'could ever have entered the Ark of Noah' (Levillier 1951). But the real breakthrough did not come until 1589 when the *Historia natural y moral de las Indias* by José da Acosta was published in Leon and, in the next six years, was translated into the major European languages. Acosta had spent five years in Peru as a missionary and during that time

had travelled as far north as Mexico. He was familiar with most of the native mammals of South America, from the big guanacos and tapirs to the small chinchillas and guinea-pigs. He knew of rheas and humming birds, of cocoa and pineapples. As a Jesuit, he believed that all animals came from the Ark but he wondered how animals could have travelled from Ararat to South America or, for that matter, to other far-flung parts of the globe 'where they say there are certain kinds of creatures that are not found in other regions.' He asked himself and his readers three questions. Were the animals of South America new creations of God? Were they taken to South America by man? Or could they have arrived by swimming when the waters covered the world? He found it difficult to accept that a 'perfect' animal like a tiger (jaguar) could have been engendered from the earth without sexual generation. And why, he asked, if all the animals were new creations were there no elephants or giraffes in South America since they would be well suited to the climate. It was equally difficult to accept that man was responsible for bringing the animals over the seas to South America: 'who would say they brought lions and tigers', he wrote, 'certainly it is a thing of mockery even to imagine.' Then it only remained to consider the third question to suppose that the

Fig. 1 Biogeographical map of 1530—P. Apium (in George 1969).

animals had arrived by swimming. But it was pre-
posterous to suppose, he argued, that so many
mammals and reptiles had successfully navigated
long distances across the ocean when even some
birds had not managed to fly to the new lands.
And so he concluded that it was probable that all
land was continuous somewhere and the animals
went from one place to another by walking across
it. Occasionally the lands heaved up, changed, so
that some animals found one place and some
another. Dispersal was the clue to distribution.

Others, like T. Fernandez and Geret de Veer
found climatic or ecological factors a satisfactory
explanation; a few, like Richard Simson, con-
templated former land connections (George
1980). But for the most part writers were content
merely to wonder why South American animals
were restricted to that continent (Burton 1624) or
why there were no native horses in America
(Browne 1635) and either to accept as sufficient
explanation the story of the Ark (de Bry 1590) or
to dismiss the question as unreal (Raleigh 1612).
There the matter rested until the middle of the
eighteenth century.

Between 1749 and 1804 the *Histoire naturelle*
of the Comte de Buffon was appearing. There had
been a considerable increase in knowledge of
plants and animals from distant parts of the
world, maps were more accurate and, furthermore, fossils had been dug up and recognized at
last for what they were. Facts had increased but
speculation on the causes of animal and plant
distribution had hardly advanced since Acosta's
day. In Volume 9 (1791) of *Histoire naturelle*
Buffon considered the faunas of South America
and Africa. The climate was similar and yet South
America had tapirs, armadillos, and agoutis,
Africa had elephants, pangolins, and jerboas.
Although the two continents had some faunal
similarities like the big cats and the monkeys even
they were not of the same species. South America
and Africa are today isolated from one another by
sea or by climatically inhospitable land, neither of
which could be crossed by their tropical animal
inhabitants. Swimming and human transport were
out of the question. Therefore, he argued, once

upon a time the animals must have had access
through the north when the climate was different
and the land was different. Natural history proves
better than geographical speculation, he continued, that the northern continents were once
joined through Kamchatka. Isolated in their new
homes, the animals of the two continents might
have changed according to the new surroundings
in which they found themselves. In Volume 14
(1766) he suggested that some of the distinctive
features of the South American and African
faunas (Figs. 2 and 3) might be accounted for by
their former contiguity and subsequent isolation
by the rise in sea-level of the surrounding oceans.
Such an event would be natural and imaginable
but it must have happened a very long time ago

Fig. 2 From Lundolf, H. (1682). *History of Ethiopia.*

and was almost impossible to reconstruct from present knowledge. Concentrating on better-known distributions where fossils were plentiful he accounted for the elephants of Africa by dispersal from the north, and the absence of elephants from South America by the barrier of high mountains on the northern border of that continent. The northern Old World was the main source of the vertebrates.

A new approach to the problems of biogeography came from the German Eberhardt Zimmermann in 1783. He tried a statistical assessment of the mammalian faunas of the world thereby recognizing similarities as well as dissimilarities in the major faunas. He counted the groups of animals that inhabited the tropics and the temperate regions; he listed and counted the mammals of different regions which he illustrated with a map. Among the five regions there was Africa with 129 groups of mammals and the Americas with 159. Per square mile, Africa had three times as many mammals as the New World. Later, he modified the regions to separate the two Americas into regions and he joined Africa with Europe. How had this regional distribution come about he asked, and reviewed the theories of his predecessors. That of Buffon seemed to him the most likely explanation, but he rejected it on the grounds that migration from the north to the south in a cooling northern climate should have put North American elephants (known from fossils) into South America as well as Eurasian elephants into Africa. He could not accept the South American mountain barrier of Buffon

Fig. 3 de Bry, T. (1644). *Amence*. (Mason B.B. 62. p132—Bodleian Library.)

because the Atlas and Altai mountains should have been a similar barrier against Africa. But he did not accept that there was once no Mediterranean Sea and free passage for monkeys and porcupines. Lack of food, he decided, was an incentive to dispersal and man might have had a considerable effect on distribution by hunting on the one hand and transport on the other. However, it was likely that animals that live together today had always lived together and had, therefore, been created on the spot.

The same contrasts between South America and Africa emerged from the study of the distribution of insects and arachnids by Pierre Latreille (1817). And although he, like Zimmermann and to some extent Buffon, was convinced that climate was an important determinant of animal distribution which united the subtropical, tropical, and temperate 'climates' of South America and Africa, yet the dissimilarities were great enough for the climates to be separated into eastern and western subtropical, tropical, and temperate. An explanation was needed of the differences in spiders, insects, and even reptiles of America and Africa. A combination of dispersal and physical barriers must be the explanation.

From this time on zonation into provinces, regions, realms, and nations was more or less taken for granted, though the authors barely questioned how such divisions had come about. Dividing the world was an interesting occupation and it might reveal patterns in the Divine Plan.

In 1826, J. C. Prichard found seven more or less latitudinal regions for mammals and joined most of Africa, India, and equatorial America into the Equatorial Region though he recognized that striking differences existed between Africa and South America. The southern extremities of America and Africa were joined into another region. Swainson (1835) objecting to Prichard's regions on the grounds that he had confused analogy and affinity, proposed five provinces. The two Americas were joined into the American province and Africa remained intact as the African Province. Swainson favoured a theory of quinary circularity. Each of the five regions was

divided into five subregions arranged in a circle. The centre of each circle was a comfortable place to live and the circumference uncomfortable. He believed neither in climate nor food, enemies nor habitat as determining global factors but only in the decree of the Almighty.

Meanwhile Alexander von Humboldt (1807) and Augustin de Candolle (1813) were advancing the knowledge of plant distribution by classifying plant growth forms and searching for an explanation of the distributional anomalies they found. The basis of their analysis was numerical: botanical arithmetic could calculate the relative incidence of particular plants in different geographical zones. Von Humboldt published an interesting profile of the vegetation of the volcanic Mount Chimborazo in Ecuador showing the stratification of plants from the top to the bottom, from the snow to the tropics. Like Candolle he believed that climate was the most important control on plant distribution. But in *Essai sur la géographie des plantes* (1805) there are echoes of Buffon. The present distribution of plants, he wrote, could indicate different states of the land in the past, of erstwhile unions of islands and 'elle annonce que la séparation de l'Afrique et de l'Amérique meridionale s'est faite avant le développement des êtres organizés.' Humboldt did not believe that animals and plants had changed nor did he believe that fossils were ancestral to living forms.

Candolle, too, was puzzled by the differences between South America with Africa but, unlike his predecessors, he was as aware of the regional similarities as of the regional dissimilarities of plants. Botanical regions could, therefore, be differentiated only by their endemics. Candolle had 20 regions which later expanded into 40. There was a minimum of four in Africa and four in South America. Plants originated at many centres and many of them stayed there. Only species that were found to have a wide distribution had needed to disperse.

There was little change until the middle of the century when Louis Agassiz (1850) proposed 11 zones, based mainly on climate, in which the tropics of Africa and South America and the temperate regions of the two continents were, once again, separate zones. The divisions grew until they reached 21 realms in Ludwig Schmarda's classification of 1853 (Fig. 4). South America was divided into four realms: Brazil the realm of edentates and broad-nosed monkeys; Peru and Chile, the realm of llamas and condors; the Pampas, the realm of viscachas and marmosets; Patagonia, the realm of guanacos and Darwin's ostriches. Africa consisted of West Africa, the realm of narrow-nosed monkeys and termites; Greater Africa, the realm of ruminants and pachyderms; and the Sahara, the realm of scorpions and African ostriches.

The system was getting out of hand: too many divisions and too many names. It was losing its usefulness. It was, therefore, an important step

forward when P. L. Sclater (1858) reduced the world to six avifaunal regions which did no correspond exactly with the continents and whose distinctiveness and scientific status were underlined by a new nomenclature. Thus South America became Region Neotropica and Africa Regio Aethiopica. Regio Neotropica formed part of Creatio Neogeana and Regio Aethiopica part of the Creatio Palaeogeana. The regions were identified more by their differences than by their similarities and each was separated from its neighbours by major physical barriers. Thus the Atlas Mountains separated the Ethiopian region from the Palaeoarctic and the lowland forests of southern Mexico separated the Neotropical from the Nearctic.

By 1859, A. R. Wallace had taken up the

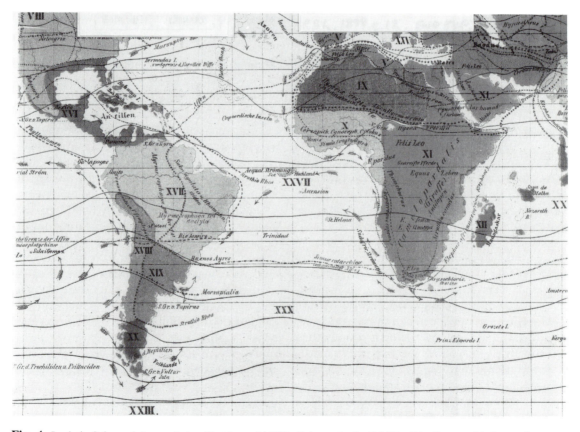

Fig. 4 Ludwig Schmarda's zonal classification of 1853. (Schmarda, L. (1853). *Die Geographische Verbreitung der Thière*.) (R.S.L. 1899. c.12. END OUP—Bodleian Library.)

challenge. He not only accepted Sclater's regions but showed that they would fit mammals and insects as well as the birds, particularly if the boundary for the Ethiopean region was the great Old World desert instead of the Atlas Mountains.

The Sclater–Wallace classification has persisted in spite of its many critics largely because, as Wallace always maintained, it could be justified by its usefulness. The zoogeographical regions have proved useful. They have been grouped in various ways to fit various organisms and authors; the north–south, or horizontal, classification of T. H. Huxley in 1868 brought in the terms Arctogaea for the north (and including Africa) and Notogaea for the south. J. A. Allen (1871) stratified the world still further, into zones distant from the North Pole by varying degrees but with South America and Africa separate. Adolf Engler (1882) studying plant distribution, followed the same line of reasoning as Huxley and separated off a southern temperate region. Names such as Allen's American Tropical, Indo-African, South American and African Temperate, or Engler's Northern Extratropical and Old Oceanic have proved cumbersome, but groupings such as Huxley's Arctogaea and Notogaea have had their adherents. It depended greatly on whether the author was a vertebrate zoologist, an invertebrate zoologist, or a botanist, and whether he was more concerned with similarities or dissimilarities. Thus, today, botanists and some zoologists unite the Ethiopian and Oriental regions into a Palaeotropical (Good 1964, Müller 1974). The southern tips of South America and South Africa are separated from the tropics by botanists but not by zoologists. But no one any longer unites South America and Africa into a unique equatorial region. Whatever their personal preferences everyone recognizes the nomenclature of Neotropical and Ethiopian and whether they agree or not with the status of those two regions use them as a starting point for discussion.

Having established a useful classification of the globe, the question immediately arose as to how to account for the similarities and dissimilarities. The Ark, climate, dispersal, special creation had all been invoked but by the mid-nineteenth century a revolution in the biological sciences gave an immediate new look to biogeography.

In 1855, Wallace had asked several seemingly unanswerable question about distribution. Peculiar groups of animals in an area were surely evidence of long isolation, he argued. Had they been formed from pre-existing more wide-ranging forms that had become extinct?

How had the biogeographical patterns been formed? That was now the important question.

With the theory of evolution by natural selection a few years later the static subject of biogeographical classification became a dynamic subject of dispersal and organic evolution, of land movements and inorganic evolution. And the comparison between the Old World and the New World loomed large in the arguments.

For if we compare, for instance, certain parts of South America with the southern continents of the Old World, we see countries closely corresponding in all their physical conditions, but with their inhabitants utterly dissimilar (Darwin 1859).

And this had come about by dispersal and evolution; evolution and interaction of animals and plants and evolution of the land. But would the living organisms move round the globe or would the land move isolating groups of plants and animals? Both Darwin and Wallace supposed that most vertebrate animals had originated in the north where they had intermittent migratory routes east and west. They had been pushed south in waves to occupy southern continents. They had mixed with the indigenous population and changed as a result of differential selection. Darwin was firmly convinced that dispersal from centres of origin (and he did not exclude southern centres entirely) was of paramount importance in plant and animal distribution and was sufficient to account for the examples of disjunct distribution which have irritated and stimulated biogeographers ever since.

Wallace, while concuring with Darwin, laid less

emphasis on dispersal, except to oceanic islands, and looked more the palaeogeographical events. He believed, as Buffon had hinted, that the study of present and past distributions would reveal past changes in the earth and these in turn would explain other distribution patterns. But Wallace was hampered by his firm belief in Lyell's uniformitarian geology. Neither Darwin nor Wallace favoured the land bridges which were rising all over the world to explain examples of disjunct distribution.

But how was Wallace to explain the porcupines and the monkeys of America and Africa? Eagler had suggested that the occurrence of similar plant types in Africa and America could best be explained by the former existence of great islands or a land mass connecting northern Brazil and West Africa across the Atlantic. This seemed to Wallace unlikely, as unlikely as the southern continent of Gondwana (Suess 1885). Did they come from the north? No monkey fossils had been found there. Could they have crossed the sea? Certainly not by swimming 'in most cases a channel of half the distance (20 miles) would prove an effectual barrier' (Wallace 1876). But a tropical raft might do: 'the fact of green trees so often having been seen erect on these rafts is most important; for they would act as a sail by which the raft might be propelled in one direction for several days in succession, and thus at least reach a shore to which a current alone would never have carried it' (Wallace 1876). But help had already arrived to solve the monkey problem. Hemprich (1820) had shown, to the satisfaction of some, that the New World and the Old World monkeys were not identical and could, therefore, have had an unknown pre-monkey common ancestor in the north. The importance of accurate classifications was coming to be realized. American porcupines inhabited the Nearctic as well as the Neotropical region. Northern origin from an unknown, undiscovered ancestor was the only possible explanation. But still it was a puzzle. Wallace wrote:

The true explanation of all such remote geographical affinities is that they date back to a time when the ancestral group of which they are common descendants had a wider or a different distribution, and they no more imply any closer connection between the distant countries the allied forms now inhabit, than does the existence of living Equidae in South Africa and extinct Equidae in the Pliocene deposits of the Pampas, imply a continent bridging the South Atlantic to allow of their easy communication. (Wallace 1880).

More detailed and sophisticated, but none the less similar thinking, persisted until the mid-twentieth century. Centres of origin, dispersal, and evolution hindered or helped by physical, climatic, and organic barriers were the key to distribution. Events since the Cretaceous were largely dominated by the idea of radial migration from the northern circumpolar land mass (Handlirsch 1913, Matthew 1915).

Meanwhile disjunct distributions were being given a quite new interpretation by Alfred Wegener. The displacement theory of continents was first put forward in a lecture in 1912 and published in book form three years later. The starting point of his theory was the striking similarity of the shapes of the coast-lines of Brazil and Africa. He collected geophysical, geological, palaeoclimatological, palaeontological, and biological arguments to show that only wandering continents could satisfactorily explain the extraordinary likeness of eastern South America and West Africa. Such a theory would account for the fossil *Glossopteris* flora of the southern continents, for the manatee, and the mesosaur reptiles. The separation of the two continents probably started in the Cretaceous, he argued, but was not complete until well on into the Tertiary. No one took much notice and Wegener's book was not translated into English until 1924. Immediately the idea was taken up by the South African geologist Alexander du Toit who, in 1927, made a 'geological comparison of South America with South Africa'. He argued that the displacement theory accounted for most of the geological puzzles: it condensed all the elements of Gondwanaland, brought the permo-carboniferous glaciations more or less together, accounted for the *Glossopteris* flora, brought triassic arid areas together and assembled mesozoic lava fields. He wrote:

Geological evidence almost entirely must decide the probability of this hypothesis, for those arguments based upon zoodistribution are incompetent to do so.

When continental drift at last became respectable through plate tectonics in the 1960s many biogeographers, following du Toit's way of thinking, considered that all their problems would soon be solved by the activities of the geologists and palaeontologists. Chironomid midges, land snails, peripatus, mudfish, nutmegs, proteas, and flightless birds of the southern continents could be accounted for by earlier widespread distribution and subsequent isolation at the break-up of Pangaea in the Jurassic and Cretaceous. But groups evolving in and after the Cretaceous still posed a problem. Modern mammals and flowering plants being comparatively recent organisms might have found the break-up of Pangaea too far advanced for them to have achieved a world-wide distribution. And the question remains. How did Turneraceae (*Strelitzia*—bird of paradise flower), characid fish (piranhas, jewel fish, tigerfish), aquatic pipid toads, and side-neck turtles, come to inhabit the Neotropical and Ethiopian regions and only those regions?

Two different methods have been used to try to answer this type of question. There are those, particularly among zoologists, who rely on a search for and interpretation of fossils (Darlington 1965, Simpson 1980, for example). There are others who maintain that only an analysis of modern faunas can give the correct answer to the question (Mayr 1944, Croizat 1958, Nelson 1969).

The palaeontologists had a lot on their side as more and more fossils were unearthed and more and more were accurately dated by radioactive decay methods. But two major disadvantages of palaeontological studies are the near impossibility of knowing whether the earliest in a fossil series is really the earliest and whether the animal or plant that it represents lived only in the place where the fossil *happened* to be found. There is

the further complication that the satisfactory classification of fossils is often difficult. There is so much uncertainty in the palaeontological interpretation that Simpson, discussing South American caviomorph rodents and New World monkeys, was forced to conclude that:

it is extremely probable that rodents and primates evolved somewhere else and reached South America by waif dispersal around the end of the Eocene. It is not clear whether they came from North America or from Africa (Simpson 1980).

Many palaeontologists have stuck to the northern origin hypothesis of Buffon and Matthew (1915) in spite of the fact that there are often no satisfactory northern fossils. And just this same scarcity of fossils has hampered the biogeographers of, for example, insects, frogs, rodents, and angiosperms. This has led even the most diehard of palaeontologists to give some consideration to deductions from modern distribution.

Various methods of assessing modern distributions have been used, among them the search or centres of origin an the calculation of similarity indices.

Centres of origin have been identified by plotting the distribution of the components of a taxon and assuming that the area of greatest diversity is the original home of the organism (de Lattin 1957, Stehli 1968, Müller 1974). The Crocodilidae are at their most diverse in South America (Stehli 1968), *Aloe* in south-east Africa (Holland 1978). But there is no direct evidence that areas of the greatest diversity of a group are the centres of origin of that group. Attemps to identify the centre by assessing relationships of the organisms can be criticized on the same grounds, particularly as there is a division of opinion among the theorists. Brundin (1972) would identify the original home as that of the most primitive of the present day organisms, others would locate it where the most specialized occur (Darlington 1970).

Similarity indices between continental faunas have been tried by many authors (Simpson 1943, Peters 1968). These have been useful stimulants to the interpretation of changing past distributions (some reptiles, for example) and also of some local recent distributions, but they do no more than say which fauna or flora is more or less like which other. For mammals, the Neotropical and Ethopian regions are very much less alike than either is to its northern neighbour. But biogeographers want to know when and where a group originated, when and where it spread, and what route it took.

The vicarian 'panbiogeography' of Croizat (1958) provided a useful shake-up. He denied the usefulness of fossils and looked for general distributional patterns of modern faunas and floras. He was strongly against divorcing phytogeography from zoogeography. Patterns of biotas, the summation of individual distribution patterns, should be sought and, once found, considered as evidence of physical events as Buffon and Wallace also thought. But vicariant biogeographers deny the possibility of dispersal across physical barriers and subsequent speciation. They deny the existence of consistent dispersal patterns across barriers. Croizat's periods of mobilism only occur when there are no barriers and they do not provide conditions for speciation.

Thus at the extreme of vicarian biogeography the similarities of African and South American plants and animals must be interpreted as the fragmented remains of a former widespread and uniform Gondwana flora and fauna (Hershkovitz 1977, for example). That the organisms must have originated somewhere is hardly considered. Alternatively, the traditional northern origin would supply a widespread flora and fauna able to disperse south when there were no physical barriers in periods of so-called mobilism (Wood 1974, 1980).

Recently, a new method has come into use, that of cladistic biogeography. In this, everything depends on accurate classification of the organisms (which hardly differentiates it from any other method). In the cladistic method of Hennig (1966) and Brundin (1972) the classification depends on the identification of sister groups by finding shared derived characters among the organisms. Discrimination between related and convergent characters, however, is let as much to personal decision as in conventional methods of classification, although applying the rule of parsimony or incompatibility can eliminate some of the possibilities (Estabrook *et al.* 1977, Morse and White 1979). Systematic relationships having been more or less established a similar procedure is used to find sister group land masses in such a way that area similarities depend on common ancestry of biotas rather than on how many taxa two areas have in common. This can lead to an interpretation of all biogeographical events in terms of land movements, of strict allopatric speciation and vicariance. But there are still alternatives in the system. The cladists may have found land connections, but they may be no nearer the identification of the original home nor the routes of spread than the palaeontologist and nineteenth century biogeographer.

An interesting and more flexible model relies on cladistic classification but leaves room for the same pattern of relatedness to be interpreted in different ways for the different components of the pattern; by vicariance for some (Patterson 1981) and by dispersal for others (Ciochon and Chiarelli 1980).

There is no doubt that vicariance inferences from modern distribution patterns stimulate the search for more accurate palaeogeographical descriptions and for better systematics. The alternative—dispersal from a centre of origin—is more difficult to test. Animals and plants may have the ability to disperse but do they, in fact, disperse and do they found potentially new species when they arrive (Mayr 1963)? There are plenty of direct examples of dispersal across water barriers (Simberloff and Wilson 1970) but not so many of speciation. The black stork *Ciconia nigra* spread from Eurasia to southern Africa after 1900 and has now established breeding groups (Voous 1960) but not yet a new species. Some authors (Gardner 1973) regard the dispersing North American opossum *Didelphis virginiana* as a species now distinct from the more southerly *D. marsupialis*. Perhaps Hawaiian

Drosophila come closest to fulfilling the requirements of dispersal and speciation (Carson 1971).

But was it possible for living organisms to disperse across the Atlantic after the end of the Cretaceous and found new taxa where they landed?

No bibliography for this chapter has been found among Wilma George's papers.

INDEX